*f*P

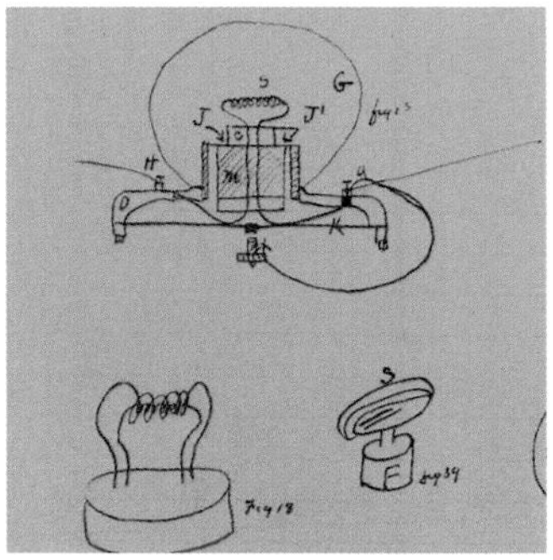

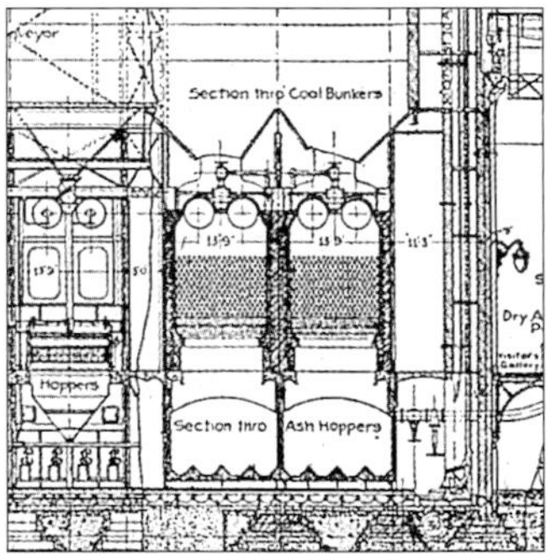
Section thro' Coal Bunkers
Hoppers
Section thro
Ash Hoppers

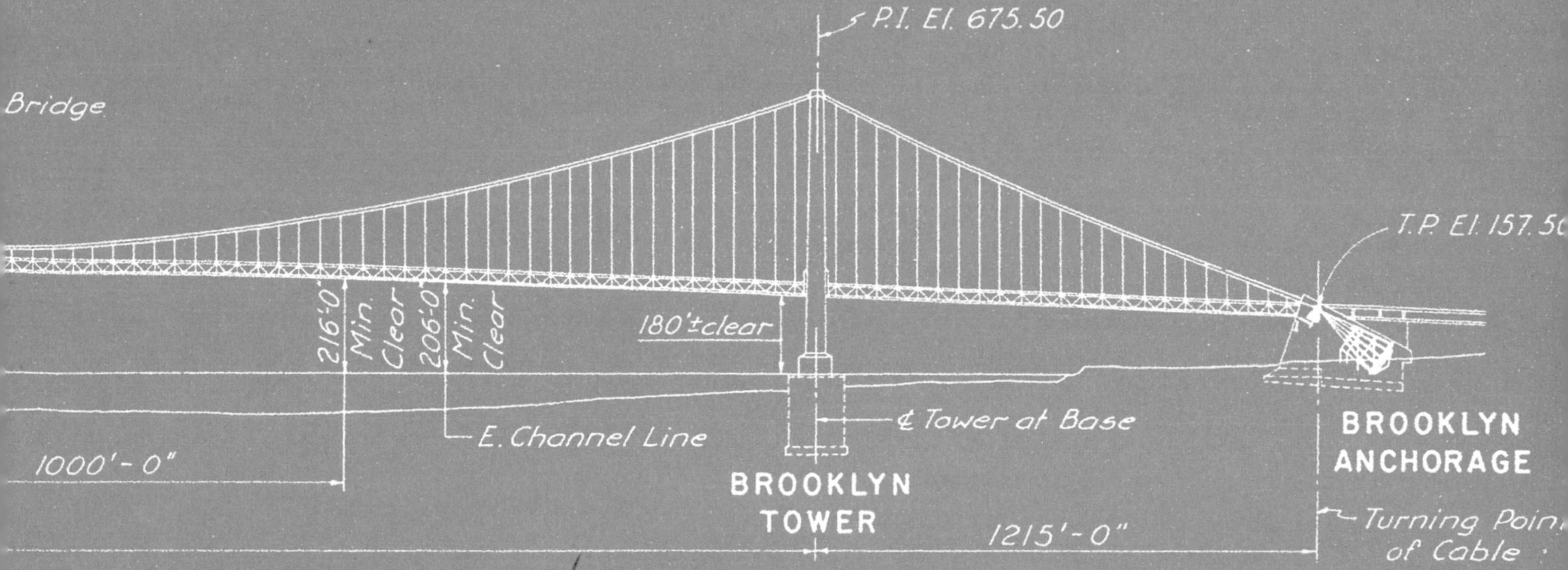

RETURN TO ROOM 131
PROPERTY OF B. GORNICK

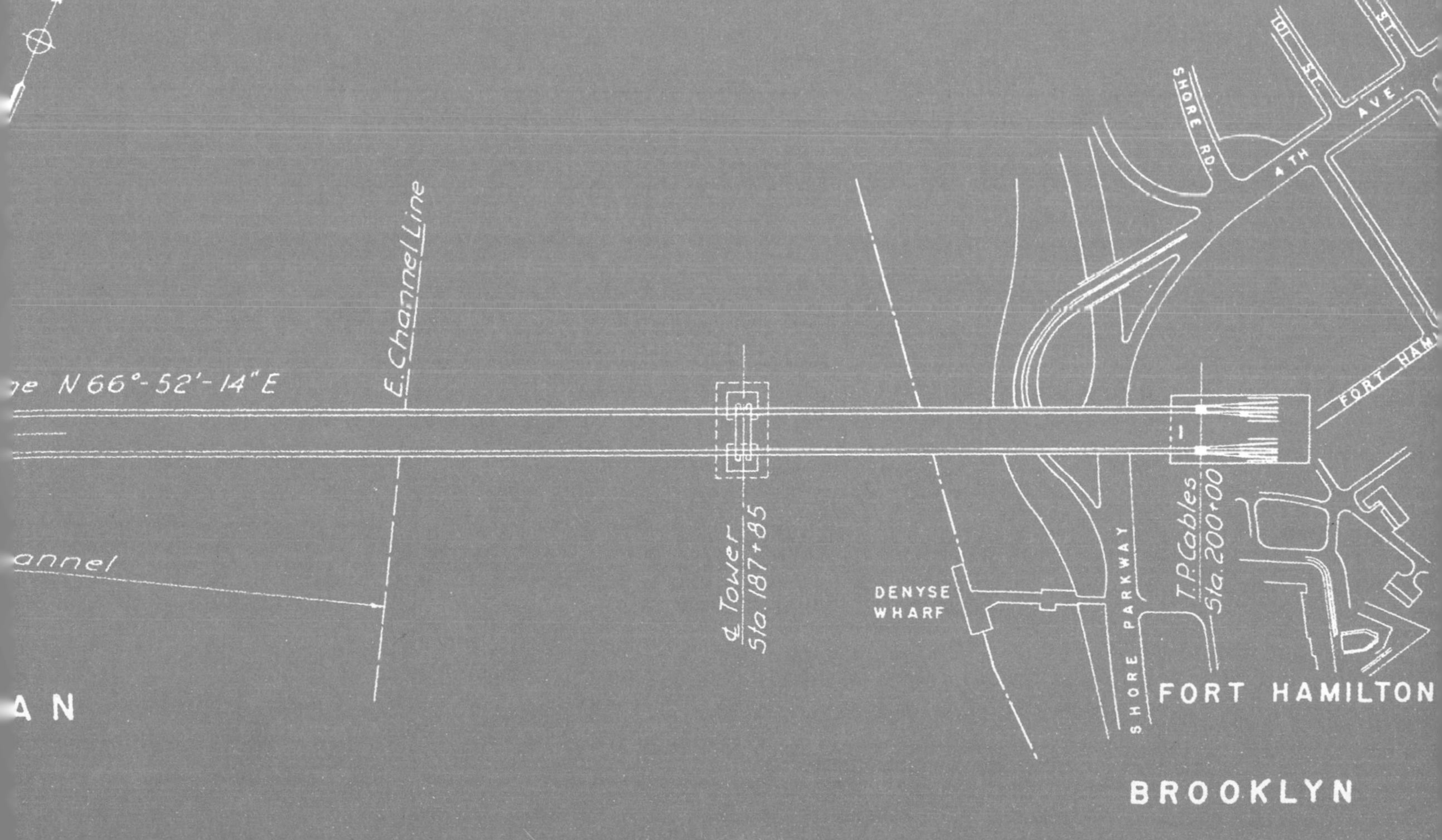

GREAT PROJECTS

The Epic Story of the Building of America, from the Taming of the Mississippi to the Invention of the Internet

JAMES TOBIN

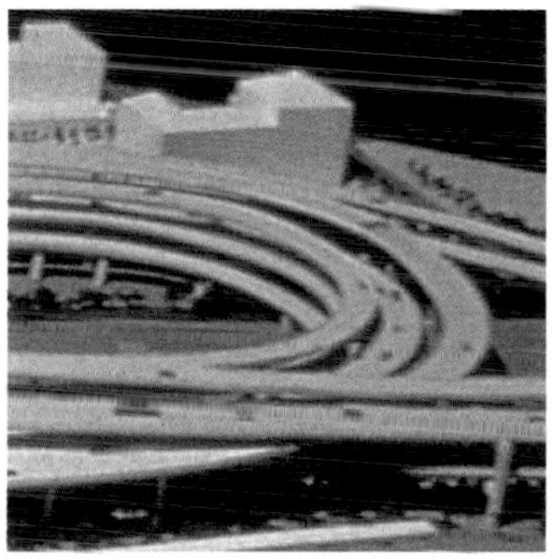

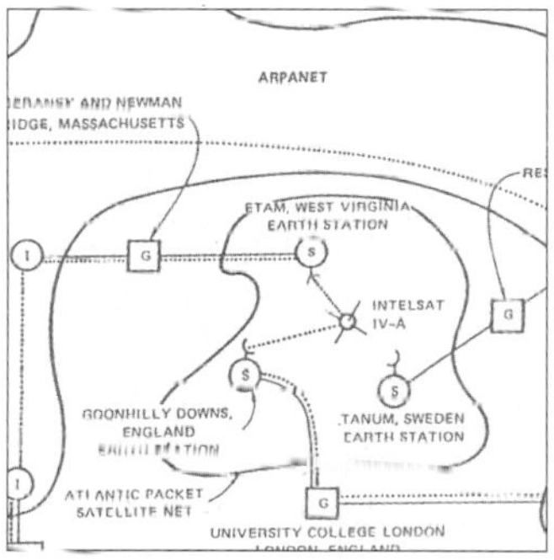

THE FREE PRESS *New York London Toronto Sydney Singapore*

fP
THE FREE PRESS
A Division of Simon & Schuster, Inc.
1230 Avenue of the Americas
New York, NY 10020

For information about special discounts for bulk purchases,
please contact Simon & Schuster Special Sales:
1-800-456-6798 or business@simonandschuster.com

Designed by Elton Robinson

Manufactured in the United States of America
10 9 8 7 6 5 4 3 2 1

Library of Congress Cataloging-in-Publication Data
Tobin, James, 1956–
Great projects: the epic story of the building of America, from the taming of the Mississippi to the invention of the Internet / James Tobin.
p. cm.
Includes bibliographical references and index.
1. Engineering—United States—History. I. Title.
TA23 .T63 2001
609.73—dc21 2001033016

ISBN 0-7432-1064-6

CONTENTS

INTRODUCTION

Anyone who has watched Frank Capra's *It's a Wonderful Life* has heard a classic expression of the American urge that lies at the heart of this book. It comes on the night when George Bailey (James Stewart) and Mary Hatch (Donna Reed) are falling in love. Following the town tradition, George makes a silent wish as he pitches a rock at the broken-down house that Mary and he soon will make their own. "What'd you wish, George?" Mary asks, and he tells her: "Well, not just one wish. A whole hatful, Mary. I know what I'm going to do tomorrow and the next day and the next year and the year after that. . . . I'm going to build things. I'm gonna build air fields. I'm gonna build skyscrapers a hundred stories high. I'm gonna build bridges a mile long."

The American experience usually is defined by reference to abstract nouns—democracy, individualism, freedom. But one concrete verb is at least as fitting—to build. That is what Americans do; George Bailey felt it in his bones. It may be unfashionable now to speak of a single, distinctive American character. But one cannot review the centuries since the American revolution without concluding that building, fashioning, on scales both small and epic, have been central to the nation's experience. Nor can one escape a sense of awe at the structures Americans have raised on the continent they claimed as their own by virtue of what they built on it. It also may be unfashionable to acknowledge the debt that Americans owe to the engineers and builders of earlier generations. Yet the civilization we take for granted would be impossible without their handiwork. The drive to put stone upon stone, creating on earth the "city upon a hill" the Puritans imagined, suffuses American history. "America is never accomplished," the poet Archibald MacLeish wrote. "America is always still to build." MacLeish meant this as a metaphor, but it was also a literal fact.

Yet historians have given short shrift to those who conceived and carried out the actual building. Just as victors tell the history of war, writers tell the history of nations. This has left engineers at a disadvantage. They tend to do their best thinking in images, not in words. They obey rules of physics the rest of us understand dimly, if at all. They work in the realm of numbers, and they believe their creations speak for themselves. Not many have written their own stories. At one point in his career, Frank Crowe, the chief engineer of Hoover Dam, was asked to write an autobiographical sketch of one thousand words. He turned in forty. Thomas Edison was called "the Wizard of Menlo Park" because most of his countrymen found his pioneering work in electrical engineering to be as incomprehensible as magic. How many of us—historians included—truly comprehend Edison's work a century later? Americans have admired their engineers from afar. But few have learned much about them.

This book is a step toward fuller appreciation of the role engineers have played in American history. It offers a history of engineering through the stories of eight great enterprises that have shaped American landscapes and expressed American dreams. The stories are grouped in four parts, each corresponding to an essential task of the engineer—to protect people from the destructive force of water while harnessing water for the enormous good it can do; to provide people with electricity, the motive force of modern life; to make great cities habitable and vital; and to create the pathways that connect place to place and person to person.

But these stories are not principally about the technical intricacies of structures and machines. They are about people, chiefly the engineers themselves—their ambitions, their battles, their genius. The stories are also about planning and politics, the processes without which great engineering projects would remain only blueprints. Not one of these projects was realized simply because an engineer had a brilliant vision, or because the project was technically feasible, or because it would benefit the public. Each had to be pushed through the sticky medium of public controversy and debate. The American tradition of public works grew up in the public arena as much as on the construction site. Indeed, the battles in the public arena were just as difficult as the work of design and construction, and they demanded at least as much persistence and courage.

The eminent American historian Daniel Boorstin once wrote that although democracy was usually described in terms of self-government, "I prefer to describe a democratic society as one which is governed by a spirit of equality and dominated by the desire to equalize, to give everything to everybody." In the United States, Boorstin said, technology has served "to democratize our daily life." The projects described here surely have expressed and accelerated that tendency. Enterprises undertaken for the sake of safety—the Mississippi levees, the Croton Aqueduct—were built for the safety of all, not only of a monied elite. Projects that aimed to enhance the quality of life—electrification, the Big Dig, the Internet—embraced more and more people as they grew. They provided tools for what the founders called "the pursuit of happiness."

As everyone knows who has watched *It's a Wonderful Life,* George Bailey never gets to build his bridges. Like most of us, he must stay at home, tending to the business of family and town. Only a few have been privileged to stand deep in Black Canyon with Frank Crowe, imagining a dam, or at the edge of the Hudson River with Othmar Ammann, imagining a bridge. But the structures they built remain, and the structures have stories to tell.

GREAT PROJECTS

1 WATER

"Water is life to any society, but if uncontrolled, it brings sorrow and ruin. If Americans meant to live beyond the Appalachians in any numbers, they would have to bring the rivers under their control."

Until a moment or two ago, by history's clock, the course of civilization has followed the course of the planet's rivers. Ancient Mesopotamia grew up along the Tigris and Euphrates. In Egypt, the means of communal life depended on the Nile; in Asia and Europe, on rivers as different as the Yangtze and the Thames. Water determined the patterns of settlement and trade. And in arid regions, rivers provided the irrigation that brought life from the soil.

The Europeans who explored North America returned with tales of inland waterways that could satisfy all of a new civilization's needs for sustenance, travel, and trade. The British forbade their colonists to settle beyond the Appalachian Mountains. But when independence was won, settlers streamed through the mountains, searching for farmland along and near the rivers. The rivers would be their highways of commerce in an era when good roads were unknown even in cities, let alone in trackless woodlands. The land rush only accelerated with the vast Louisiana Purchase of 1803. By that year, flatboats and keelboats heavy with frontier produce already filled the western rivers, all of them heading downstream toward the port of New Orleans. When Robert Fulton built a practical river steamboat in 1807, two-way river traffic promised a spreading bonanza of inland trade.

But the rivers would not submit to this new regime. Americans no sooner had followed in the explorers' tracks than they learned that the rivers moved and changed by their own rhythms, with no regard for people who wanted to stay put along their banks. Water is life to any society, but if uncontrolled, it brings sorrow and ruin. If Americans meant to live beyond the Appalachians in any numbers, they would have to bring the rivers under their control. They turned to their engineers for help as an invaded people turns to its soldiers.

With the experience of wilderness fresh in their collective memory, Americans often spoke of natural phenomena as their enemies. Engineers came to see Nature differently. They learned to respect Nature as a magnificent adversary, even as a potential partner. They sought to enlist Nature's aid whenever possible and to engage it in outright battle only when necessary. They came to understand the human encounter with Nature not as a series of triumphs but as a process of give and take, of adjusting always to the unforeseen and the unforeseeable. Mastery, the engineers learned, would always elude them.

They learned these lessons best in the valleys of two rivers. One flowed through the green and prosperous heart of the continent, home to millions, the other through lost and silent canyons of stone.

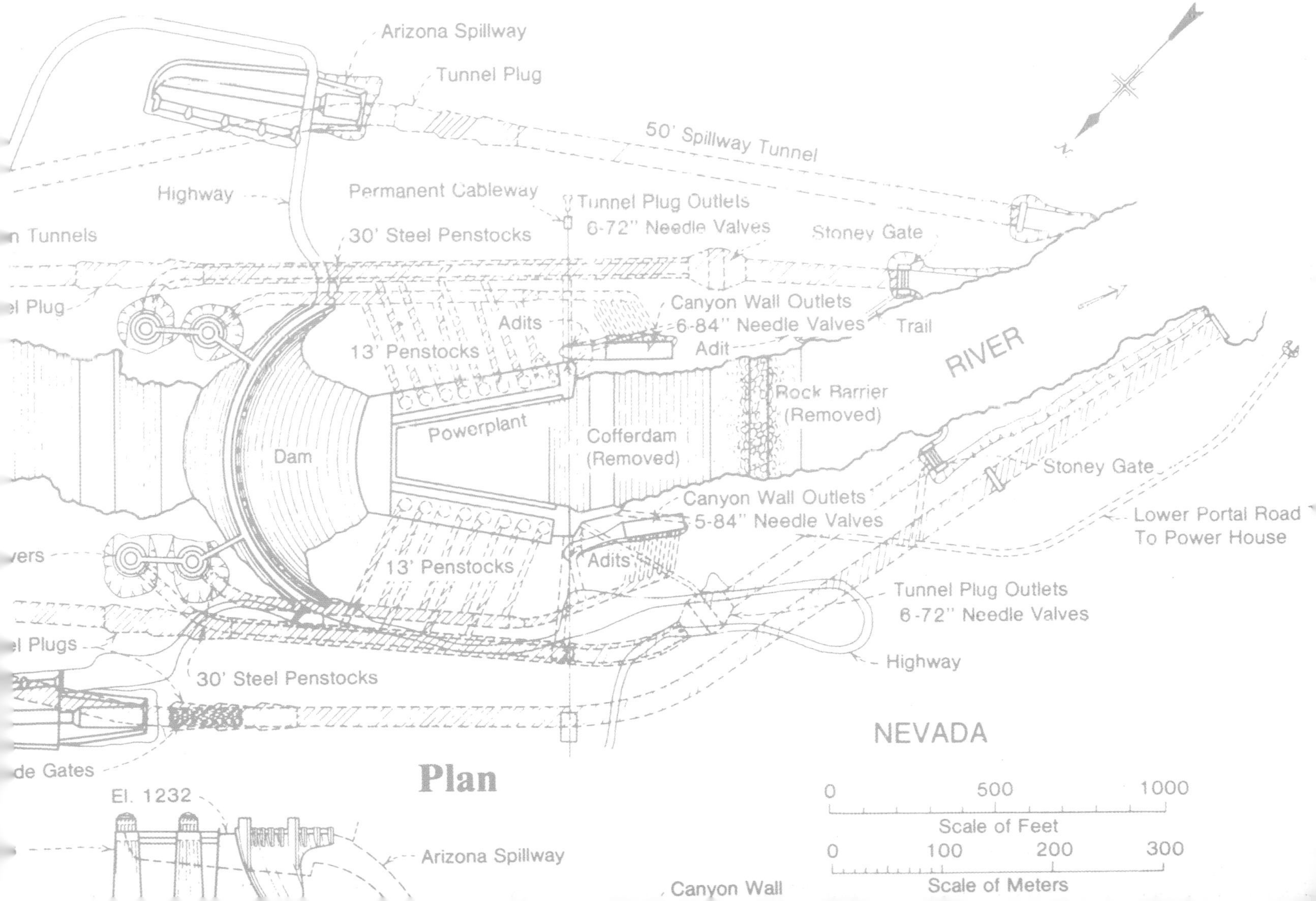

CHAPTER ONE
THE LOWER MISSISSIPPI

. . . [T]en thousand River Commissions . . . cannot tame that lawless stream . . . cannot say to it, "Go here," or "Go there," and make it obey.

– *Mark Twain*

. . . [E]very atom that moves onward in the river . . . is controlled by laws as fixed and certain as those which direct the majestic march of the heavenly spheres. . . . The engineer needs only to be assured that he does not ignore the existence of any of these laws, to feel positively certain of the result he aims at.

– *James Buchanan Eads*

The river man watched as thunderheads blew in from the west, throwing the valley of the Ohio in purple shade. Soon rain fell. Now after months of drought the river would rise, and Henry Shreve could go down the great Ohio to the greater river in the West, and begin the job no one thought he could do.

He was forty-four in the summer of 1829, thick in the shoulders and upper arms from his early years of excruciating labor, but thickening at the belt now, too. His brown hair was wavy, his eyes gray. He was given to long contemplative silences interspersed with bursts of creative energy. Shreve had first seen the Mississippi as a boy, in 1799. By that year Americans west of the Alleghenies were already sending prodigious amounts of farm produce, whiskey, and furs down the river to New Orleans by raft, barge, and keelboat. Gravity supplied all the power needed for the downstream trip. But to go upstream, crews had no choice but to pull and push the boats themselves. Tapping the potential of the inner continent demanded that someone figure a way to drive boats *up* the river as well as down.

Then in 1816 Henry Shreve designed a river-worthy steamboat and proved it could beat the current. That craft, with its shallow draft and high-built decks, revolutionized western transportation and made Shreve's name famous. He married happily, built and ran more steamboats, made money.

Yet for nearly ten years he devoted much of his time and considerable mental powers to an engineering problem that nearly all the river men of his generation believed insoluble. Why would he undertake this new errand in the wilderness—a fool's errand, nearly everyone said, even for a man of Shreve's achievements?

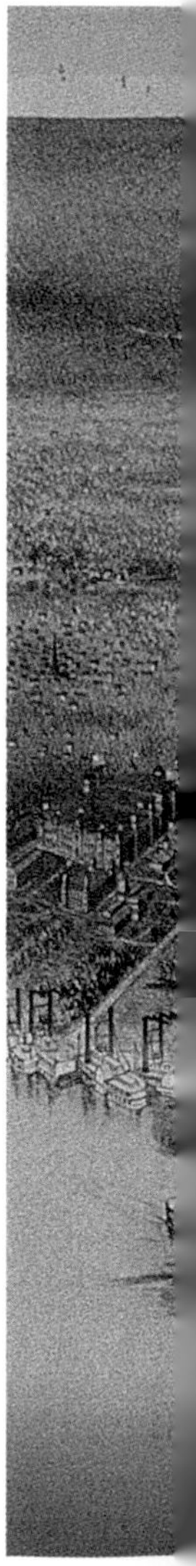

Henry Shreve already had pioneered upriver navigation of the western waters when Washington called for help in protecting steamboats from thousands of waterborne snags. Shreve's invention, the snagboat, appears in the background.
"Captain Henry Miller Shreve (1785–1851) Clearing the Great Raft from Red River, 1833–1838," by Lloyd Hawthorne.

"The City of New Orleans and the Mississippi River, Lake Ponchartrain in Distance.."

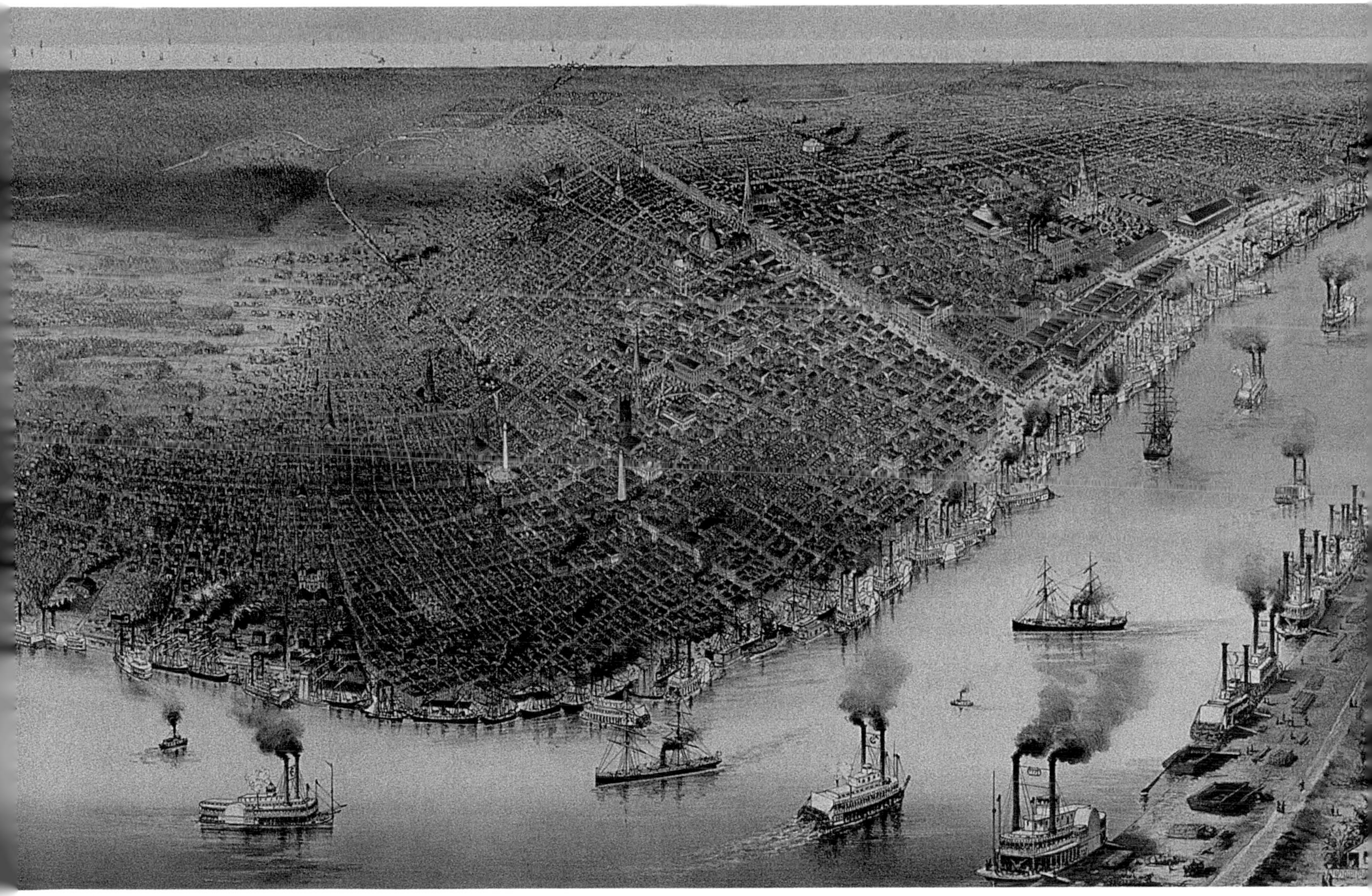

The Mississippi River.

Shreve left no record of his deepest motives. But surely he was driven by the same westward imperative that had driven his father, a veteran of Valley Forge, to carve a wilderness farm from the Allegheny forest, and his older half-brother to seek his fortune on the western rivers. The revolutionary generation had made the new land their own. Now Shreve's generation would pry the land open to extract the prizes that lay within, and they quickly learned that the richest prizes lay in the valley of the Mississippi River. There they found agricultural land as fine as any in the world and a waterway that connected the nation's midsection with the ocean trade routes. But using the river for transport would not be easy. And the job of claiming and safeguarding the land—and the works of civilization they built on it—would be harder still. For of all the natural adversaries greeting Americans on the new continent, the lower Mississippi River was the most dangerous and the least tractable.

"TEETH OF THE RIVER"

Long before Henry Shreve first boarded a flatboat, his countrymen realized the Mississippi and its tributaries comprised one of the world's great systems of navigable water. The main stream is really two rivers—the upper Mississippi, which flows from snow-choked headwaters near Canada to the center of the continent; and the broader, more baffling lower Mississippi, which carries the waters of the upper river, the Missouri, and the Ohio to the sultry salt marshes of the Gulf of Mexico. Together, the Mississippi and all its tributaries form a fifteen-thousand-mile web of waterways stretching from present-day New York to Montana, through nearly half the lands of the continental forty-eight states. Its shallow valley is so broad, a traveler once said, that "a native tribe from its eastern rim of the Alleghenies might spend a generation migrating through it before sighting the Rockies that walled it on the west."

The river came to be called "Father of Waters," but in geological age it was barely a toddler, and a defiant one at that.

It sprang into being as the glaciers of the last Ice Age lumbered northward, trailing swampy streams that became the Ohio, the Missouri, and a hundred others. Where these streams joined and ran south to the ocean, they dug a groove through the soft slab of clay and sand that became the American South. Every year, every droplet of spring rain that fell upon much of the continent ran downhill in the direction of that groove—the Mississippi's riverbed—each rill carrying a microscopic load of organic material and sand. Coursing southward, the water scoured more soft earth from the banks and bottom and whisked it along. Below the points where the Missouri and Ohio joined the master stream, uncounted tons of dirt swirled in the water like muddy clouds. (Farmers would say it was too thick to drink and too thin to farm.) The water and its burden of sediment determined the river's ever-shifting geography, and the geography challenged the ingenuity of the Americans who settled on the banks.

They soon learned that the river, like a creature of myth, possessed the power to change its shape, and did so again and again. Wherever the current slows, especially on the inside curve of a bend, sediment falls to the bottom. There a sandbar forms. The bar nudges the oncoming current toward the opposite bank. That bank crumbles under the current's pressure, sending more sediment downstream, where it collects elsewhere as a new sandbar. Mile after mile, century after century, the river undulates in a crazy squiggle of curves and horseshoes.

This process of geological transformation handed Henry Shreve his great challenge. In his era the trees of the original North American forest still stood guard up and down the Mississippi. Wherever the current assaulted the banks, these old giants lost their footing, slid into the river, and floated downstream as dangerous snags. Some blundered into sandbars or islands, catching other snags and blocking the watercourse. Some wandered free, their bobbing limbs beckoning to rivermen in slow, sinister gestures of invitation. (These were called "sawyers," for the way they appeared to saw the surface.) The deadliest snags were called planters—trees whose roots caught in the silt floor and stuck fast. More silt would pile against the roots, embedding the tree, its shaft pointing to the surface like a battering ram. When a steamboat struck a planter, the murderous effects were instantaneous—"a sudden wrench, the rush of sucking water, a clanging of bells, terrified screams, and the current would sweep over another tragedy."

The snags had been falling into the Mississippi since the Mississippi began. The early French settlers at New Orleans had called them *chicots*—"teeth of the river." By the 1820s, American surveyors counted 50,000 of them. And those were only the ones they could see. As steam traffic mounted, so did the damage and loss of life. Fear of snags slowed commerce. Military men, including Generals Andrew Jackson and William Henry Harrison, said snags impeded their mobility in the War of 1812 and the Indian campaigns. Pleas for federal action rose. In the summer of 1818, the influential *Niles' Register* reported: "Three steam boats have been lost in five months . . . in consequence of running foul of great trunks of trees. . . . Will not the increased navigation of this mighty stream soon justify an attempt to clear it of such serious incumbrances—or is it practicable to do it?"

That was the question: Did any effort to clear the Mississippi of 50,000 snags even stand a chance?

Henry Shreve believed so. As an owner of steamboats, he found the sodden driftwood hindering his every trip, costing him money and threatening the safety of passengers and crews. The great logs became a personal challenge, even a mild obsession. At home and on the river, he sketched plans for a strange new river craft.

Then, in 1824, the glowering South Carolina statesman John C. Calhoun, as secretary of war, invited river men to submit ideas for snag clearing. Shreve offered his plan for a twin-hulled steamer rigged with an apparatus for sawing off snags underwater and hauling them out with a manually operated windlass. Calhoun never responded. Yet no other river man produced a practical idea.

Soon a second plea issued from Washington. This time Shreve kept quiet, and the War Department hired John Bruce, a Kentuckian like Shreve, to clear all snags from the Ohio and the lower Mississippi for $65,000. Bruce tried his own twin-hulled design, failed, then attacked the Ohio snags with teams of men armed only with saws and chains. After two years they had barely begun on the Ohio, let alone the Mississippi, and Bruce's money was gone.

Under increasing pressure from the West, the new secretary of war, James Barbour, cast about for some new plan. He spoke to Calhoun, now John Quincy Adams's vice president, who recalled Shreve's ideas. Though Shreve was well-known as an Andrew Jackson man—and thus a natural enemy of Adams, who had defeated Jackson for the presidency in 1824—the administration set aside its partisan reservations and hired Shreve as superintendent of western river improvements under the army engineers, at a salary of six dollars per day.

Shreve had to send a stream of beseeching letters before Washington would authorize construction of his snagboat for $12,000. Finally, after he reported success with a model—"The machine is beautifully simple and most powerful in its operation and produces the effect intended in the most admirable manner"—the army agreed to pay, but told Shreve that if his experiment failed, he would bear the cost himself.

At New Albany, Indiana, a few miles down the Ohio River from Louisville, Shreve's idea took on solid form. Two steamboat hulls 125 feet long lay side by side, each with a paddle wheel on its outer edge. Beams connecting the hulls supported an overhead pulley-and-windlass apparatus for lifting snags from the water. At the bow was a wedge-shaped beam made of heavy timber and sheathed in iron. This was the weapon Shreve would thrust at snags. He named the boat *Heliopolis.*

Few on the rivers believed it would succeed. Some even stooped to political sabotage. A fellow river captain wrote the War Department: "It is said that the present Superintendent has it in contemplation to construct a large and powerful steamboat, for the purpose of cutting out the snags, and pulling them out by the force of steam. Now, those projects are only calculated to get through the appropriation, without anything like the object contemplated. All machinery, whatever, whether used by lever or steam power is considered by persons who are well-acquainted with the Mississippi river navigation, as a useless expenditure of time and money." But other captains signed petitions on Shreve's behalf. They wanted action against the rotting obstacles, however unlikely the result.

In the spring of 1829, the *Heliopolis* was finished. But a drought had drained the Ohio to its lowest level in years, keeping Shreve in port. He waited. In mid-August heavy rains fell and the river swelled. The *Heliopolis* made its way down the Ohio, then the Mississippi, to Plum Point, Tennessee, a place studded with snags, one of the most dangerous places on the entire river. Shreve arrived at midnight on August 19.

In the morning, as he got ready, other boats stood at a distance, their crews expecting the first collision with a snag to jolt the snagboat's boilers into explosive fury. Shreve chose a planter and turned the boat toward it. The *Heliopolis* gathered speed and closed in. At full steam the iron wedge crunched into timber. The tree snapped. Windlasses whirled and chains clanked, and the planter emerged, dripping, from the brown depths. The trunk had broken several feet below the surface of the sandy bottom. In moments, Shreve's men sawed it into small pieces.

Eleven hours later, the entire Plum Point channel was clear.

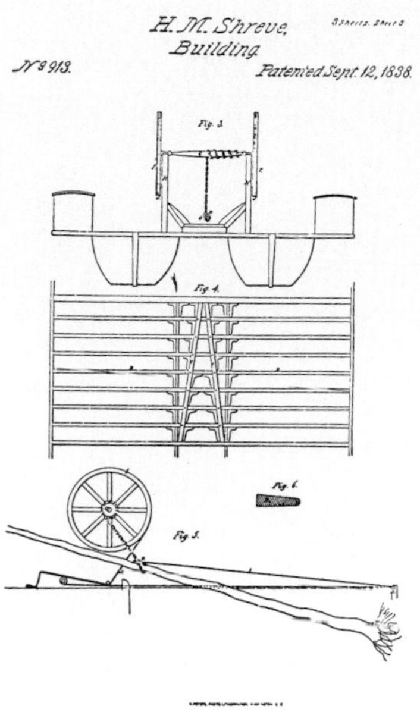

U.S. Patent Office diagrams of Henry Shreve's design for the *Heliopolis*, the first of many government snagboats constructed to make the Mississippi safer for steamboats.

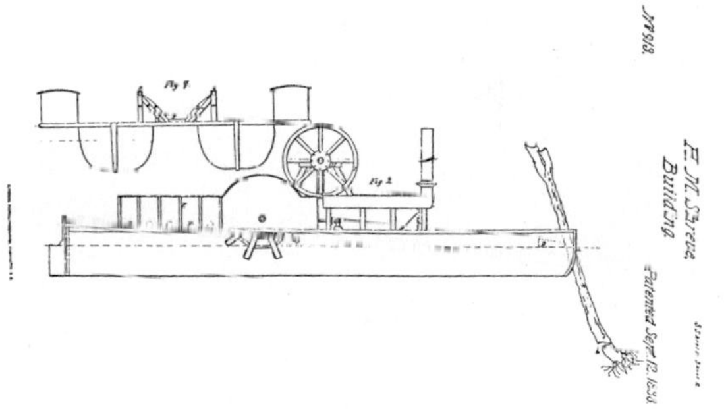

Government snagboat *Horatio G. Wright* at work clearing the river.

"I am proud to say," Shreve soon wrote the chief of engineers, "that the performance far exceeded my most sanguine expectations."

That fall Shreve roamed the western rivers. By 1830 his triumph was known throughout the country. "Capt. Shreve has perfectly succeeded in rendering about 300 miles of river as harmless as a mill-pond," one newspaper reported. A second snag boat was commissioned, then a third and fourth, all virtually identical to the *Heliopolis.* Soon the Ohio and Mississippi were all but free of snags, then the Arkansas. Finally, in his most remarkable achievement, Shreve cut through the immense log blockage of the Red River in Louisiana known as the Great Raft, 150 miles long, all the way to a place that would soon be called Shreveport. From Texas to Pennsylvania, the rivers were clear.

Of course, even a clear river is only as safe as its channel. Shreve's ingenious boats had made the river safe during normal seasons. But the Mississippi is capable of frightening departures from the normal.

AN ENDLESS WALL

[Y]ou hardly ever see the river, but the levee is always close by, a great green serpent running through woods, swamps, and farms, with towns nestling close to its slopes. The levee is unobtrusive, since its slope is green and gradual, but in fact it is immense—higher and longer than the Great Wall of China, very likely the biggest thing that man has ever made. . . . It was the principal human response to the titanic power of the great river.

—Alan Lomax, *The Land Where the Blues Began*

A band of mosquito-ridden Spanish soldiers were the first people of European origin to see the Mississippi in flood. In 1543 they crossed a bewildering inland sea that drowned all but the tallest trees, and they paddled like hell to leave it behind.

What they saw was simply the lower Mississippi being its natural self. The river rises with the rains every spring and often spills out on the land. Through the millennia since the last Ice Age, the spreading floodwaters have dropped layer upon layer of sediment, the heaviest sediment nearest the river, thus creating a rich cake of soil up and down the banks. The land sloped away from these natural mounds and turned swampy, but under the swamps the land was highly fertile, too.

The French, who came a century and a half after the first Spanish explorers departed, weren't so easily scared off. They built a town, New Orleans, on the broad mound of sediment between the river and Lake Pontchartrain. To keep the spring high water out of their streets, they constructed a dike along the river. They called it a *levée,* from the verb *lever,* to raise. By 1727 it was three feet high, eighteen feet across the top, and a mile long. If they had known what Herculean labors they were setting in motion by building that little wall, they might have packed their *bateaux* and followed the Spanish. For one levee inevitably begat another levee, and so on ad infinitum.

This the Americans discovered soon after President Thomas Jefferson bought the vast, unexplored territory of Louisiana—essentially the entire western drainage basin of the Mississippi—from the French in 1803. They learned that if you built a three foot-high levee on the east bank of the river, while on the west bank there was no levee, then the next flood crest would overflow the west bank. So the people on the west bank would build a six-foot levee and push the next flood back to you. The logic became obvious: Wherever people proposed to live year round, there must be levees.

At first, planters along the river bore the expense of building the mounds on their own land. But as it became clear that the levee was only as strong as its weakest section, and as costs mounted, levee districts were organized and given the power to raise taxes that would spread the financial burden among all those who benefited from the levee's protection. In some years, such as 1828 and 1844, high waters broke through the walls to flood farms and villages. But in most years the levees worked. They protected existing farmland and made it possible to drain and plant new farmland. More people came, and they depended on the levee, too. And so it went on both sides of the river, foot by foot and mile by mile.

By 1850 the costs were crushing. Under pressure, Washington bequeathed millions of acres of federal swamplands to the riverine states, which then sold the land to raise money to build more and

Mule skinners and mules were the essential levee builders for many decades.

better levees, which meant that still more swamps could be drained to make farms, and still more people depended on the levees.

As the levees became indispensable, their construction became more sophisticated. The early levee builders had simply piled dirt on the riverbank in long mounds filled with stumps and logs. A major high water could easily shove many of these mounds aside. Or the water, soaking the levee, would cause the stumps and logs to rot, leaving cavities that collapsed under pressure. Gradually, civil engineers learned the river's destructive tricks and how to combat them, and by 1880 they had developed a standard levee design.

First, workers would clear all debris from a wide ribbon of space, usually well back from the river itself. Down the middle of this space they dug a ditch three feet deep and three feet wide. They refilled the ditch with fresh dirt, then piled a mound of dirt on top of the dirt in the ditch. By this process they created one solid mass of material, some below ground and some above, with the ditch dirt anchoring the levee to the natural ground as the whole mass compacted and hardened. The river couldn't so easily push this anchored mass aside.

The levee was to be 3 feet taller than the presumed high-water maximum. The base at ground level was to be 5 to 7 feet across for every foot of vertical height. The levee should be as wide at the top as it was high. Thus, as the levee grew taller, it grew much broader. A levee just 7 feet high would be 7 feet wide at the top and 35 to 50 feet across at the base. After another ditch was dug to lure away the water that inevitably oozed through the structure, workers seeded the whole mound with Bermuda grass, which in time formed a tough, soddy seal that withstood high water far longer than exposed ground.

These rough engineering works were hardly perfect. Men had to patrol the levees to spot trouble, especially in spring. They looked for animals, even crawfish, digging tunnels through the mounds. They looked for sand boils, the little geysers that erupted when the river's colossal weight pushed jets of water under or through the levee,

"Scrapers at Work near Beulah, Mississippi, 4-16-1913," by F. A. Rosselle.

eroding it from within. And in flood times, the guards watched the river for men approaching by boat with explosives. If they blew a hole in your levee, inviting the river to destroy your land, they might save their own.

And the river did increasingly hold the power of destruction. For the levees were built on a paradox. The higher they rose, and the more continuous their line on both sides of the river, the more protection they afforded—as long as they did not break. If they broke, their great height actually *increased* the risk that the ensuing floods would drown and destroy.

Why? Because untold millions of tons of spring floodwaters now stood much higher—nearly forty feet high in some places—than the

A "crevasse," or break in a levee, could unleash anything from a rivulet to a devastating wall of water. "A 'Crevasse' on the Mississippi," by Alfred Waud, 1874.

ground on which the levees were built. When that much water stood so high above the land and then tore a hole, or "crevasse," in the levee, simple gravity would force the water to fall on the land with a force akin to Niagara Falls.

Engineers on the River

Whenever the levees failed, the people of the lower Mississippi begged the nation's engineers for solutions. But to tame the river the engineers had to understand it, and its behavior was nothing short of bizarre.

As John M. Barry wrote in *Rising Tide,* his comprehensive study of the control of the lower Mississippi and the 1927 flood, the river "moves south in layers and whorls, like an uncoiling rope made up of a multitude of discrete fibers, each one following an independent and unpredictable path, each one separately and together capable of snapping like a whip. It never has one current, one velocity. Even when the river is not in flood, one can sometimes see the surface in one spot one to two feet higher than the surface close by, while the water swirls about, as if trying to devour itself. Eddies of gigantic dimensions can develop, sometimes accompanied by great spiraling holes in the water." Where the watercourse bends, "the collision of river and earth . . . creates tremendous turbulence: currents can drive straight down to the bottom of the river, sucking at whatever lies on the surface, scouring out holes often several hundred feet deep."

The river's variables were many and they changed constantly. Heavy rainfall or collapsing banks could shatter months of careful observations. The shape of the riverbed's underwater cross section varied wildly from place to place. Tidal waters entering from the Gulf of Mexico affected the river's movements for many miles upriver. At flood stage the river behaved entirely differently than at low water. Levees that held a large high water might collapse under a lesser one. And hard as it was for the engineers to detect any predictability in the water itself, it was perhaps even harder to perceive rhyme or reason in the behavior of sediment—how it moved at various depths, why it fell to the bottom where it did, how it affected the speed and direction of the current. Whenever one set of conditions seemed about to yield to sustained study and analysis, the river would blast those conditions to hell and make something wholly new and unknown.

In the mid-1800s, two schools of thought gradually emerged about what to do. Given the scarcity of dependable fact, it's hardly surprising that they seemed to be diametrical opposites.

One school argued that the river could be enlisted to dig itself a deeper bed, which would hold more water and thus prevent floods. The theory had its roots in the valley of the Po River in northern Italy. There, an engineer named Domenico Guglielmini (1655–1710) had theorized that if you built walls to confine an alluvial river such as the Po, you would force the river to rise. A higher river has a steeper slope to its outlet, and a steeper slope makes the river run faster. A faster river scrapes up more sediment from its bed. In short, a river confined within levees would dredge a deeper bed for itself, and thus carry more water in times of flood. The "levees-only" theory, as it was known, scored some successes in controlling floods in Europe, so it gained adherents in the Mississippi valley. Some even argued for closing off the Mississippi's natural outlets to increase the river's scouring power. But they had fierce opponents who believed levees were little more than a necessary evil.

The idea of the opponents of "levees-only" can be expressed by analogy to a kitchen sink. When someone closes the drain and turns on the faucet, water climbs up the walls of the sink. If no one opens the drain or turns off the faucet, the water overflows onto the floor. That is the nature of rivers, too, but no one can turn off the Mississippi. If you can't turn off the faucet and you don't want the floor to get wet, it doesn't make sense to build up the walls of the sink. It makes sense to open the drain.

To keep the Mississippi from flooding, the opponents of "levees-only" said, its drains must be kept open. Dredges should be employed to keep silt out of the river's natural outlets, such as the Atchafalaya River and other Louisiana bayous, so they could better absorb floodwaters. Reservoirs should be built on tributaries to hold floodwaters out of the main river. And most important, artificial outlets should be built to allow floodwaters to spill off into alternative routes to the Gulf. These ideas comprised the "outlets" theory of flood control.

To resolve the argument, Congress decided in 1850 that a major federal survey was needed to wrest the secrets of flood control from the lower Mississippi once and for all. Because this was to be one of the great scientific undertakings of the era, leaders of the Army Corps of Engineers assumed the survey would be theirs. After all, the profession of engineering in America had been virtually synonymous with the corps for half a century; few influential engineers had been trained outside West Point. But as the industrial revolution spread, engineers outside the military were rising in numbers and influence, and they demanded the Mississippi survey be assigned to one of their own. So President Millard Fillmore simply split the $50,000 appropriation in half and told the army to do one survey and the civilians another.

The civilian effort fell to Charles Ellet. One of the first American engineers to spurn West Point, Ellet took his training at the more scientific École des Ponts et Chaussées, in Paris. Handsome, brilliant, and brave to the point of foolhardiness—he was the first person ever to cross the gorge at Niagara Falls—Ellet commanded respect. He compiled his report in under two years. Short on data but long on vision, the report called levees-only "a delusive hope, and most dangerous to indulge, because it encourages a false security. . . . The water is supplied by nature, but its *height* is increased by man. *This cause is the extension of the levees.*" Instead, Ellet called for the dredging of natural outlets and the construction of artificial ones. For a few years in the 1850s, his work was celebrated as the authoritative statement on the Mississippi.

His rival in the corps of engineers was Andrew Atkinson Humphreys, a driven and driving man who, despite his army background, approached the study of the river with far greater scientific tenacity than Ellet. This meant that he took longer than Ellet. He chased every last fact, tested every theory. He became so consumed by his task that he suffered a nervous breakdown, lost his funding, and had to set the work aside for several years. But in the summer of 1861, amid the first clashes of Union and Confederate armies, Humphreys completed his survey. It became the most influential scientific study of the Mississippi ever written.

Humphreys was no more persuaded by the levees-only theorists than Ellet had been. Observable facts proved Guglielmini's idea that levees would lead to a deeper river "totally erroneous," Humphreys said, and that closing natural outlets would be "disastrous."

But Humphreys was equally dismissive of the outlets theory. He said flood reservoirs on the Mississippi's tributaries would not work. He said artificial outlets, though they would indeed lower the river, would be too expensive relative to the value of the property they would protect. And, he said, outlets posed a grave, heretofore unforeseen danger. In a great flood, the main stream of the Mississippi might shift entirely to the new outlet, leaving the old channel, with all its human settlements, an isolated backwater. To guard against floods, he said, the nation would have to make do with strong levees. They would not make the river deeper, but they "may be relied upon for protecting all the alluvial bottom lands liable to inundation."

The Civil War distracted all attention from river hydrology. But it decided the winner of the great river debate. Charles Ellet was killed on a Union battering ram. Andrew Humphreys not only survived the war but achieved renown as a field commander at Fredericksburg and Gettysburg. To reward his war service and his river survey, he was named chief engineer of the army in 1866.

Ellet's case for outlets faded away, and Humphreys's views prevailed. Unfortunately, the nuances of his massive report were not entirely understood. People understood him when he said levees would do the job. But many overlooked his dismissal of the key idea of levees-only—that levees would make the river deeper. So the levees-only theory lingered in the public mind. And in the 1870s it appeared to draw new vitality from the extraordinary achievement of a brilliant, disciplined, austere man who was regarded as the nation's greatest engineer, James Buchanan Eads.

Andrew A. Humphreys, U.S. Army Corps of Engineers, whose confidence in the ability of levees to protect against floods led to the corps's adoption of the "levees-only" policy to control Mississippi River floods.

James Buchanan Eads understood the Mississippi better than any other man of the nineteenth century. His mastery was put to its ultimate test at the river's mouth, where giant sandbars threatened to close the river to oceangoing commerce.

Mr. Eads at the Passes

Eads had learned the Mississippi's strange ways as no man ever had. In 1838, as a steamboat man barely out of his teens, he had spotted opportunity in the hundreds of shipwrecks sent to the river bottom by fires, boiler explosions, and Henry Shreve's snags. Some said the river's crazy currents and shifting floor made it impossible to salvage the wrecks. Eads disagreed. He built a salvage vessel much like Shreve's snagboats. To locate lost cargos in the silt-laden water, he made some three hundred descents to the bottom in dangerous diving bells. About one dive he wrote:

"The sand was drifting like a dense snowstorm at the bottom. . . . At sixty-five feet below the surface I found the bed of the river, for at least three feet in depth, a moving mass and so unstable that, in endeavoring to find a footing on it beneath my bell, my feet penetrated through it until I could feel, although standing erect, the sand rushing past my hands, driven by a current apparently as rapid as that on the surface. I could discover the sand in motion at least two feet below the surface of the bottom, and moving with a velocity diminishing in proportion to its depth."

These intimate explorations provided all the data that Eads's acute mind needed to master the river in various ways. He made a fortune in salvage. During the Civil War he constructed ironclad warships for the Union navy. After the war, at St. Louis, he built the first bridge over the Mississippi, proving the value of the new experimental metal, steel.

Even before his bridge was complete, Eads, now fifty-five years old, tackled a still greater challenge, this one in the deepest South, where the Mississippi splits into three short branches, or "passes," and flows into the Gulf of Mexico. What he accomplished there had nothing to do directly with flood control. But it would profoundly affect the flood control debate.

Here at the river's mouth lay muddy lumps of geological goo from every acre of land between the Pennsylvania coal fields and the Yellowstone wilderness. These great sandbars were growing noticeably larger by the year, slowing the passage of ships in and out of the river and threatening to block them entirely.

The sandbars were the natural product of the river going about its business. Every day the current dropped more sediment at the mouth. In time, when there was more sediment than water, the river in its nonchalance would simply spill off in some new route to the Gulf. It had done so for eons.

But Americans had committed themselves to *this* route, lining it with cities and farms and port facilities. They couldn't wait for a century as the Mississippi slowly changed its mind. The sandbars threatened the prosperity of the entire valley, even the nation. In 1859, a visiting business delegation counted three ships stuck on sandbars, thirty-five ships waiting upstream to get into the Gulf, and seventeen waiting in the Gulf to get into the river. Action was essential.

The corps of engineers failed in one attempt after another to get rid of the bars. In desperation, the New Orleans Chamber of Commerce demanded that a canal be dug to bypass the blockage altogether. The corps agreed. Ships still would have to stand in line, because the canal would be wide enough for only one ship at a time, but at eighteen feet it would be deep enough for most ships. It would cost $13 million. The canal's champion was the corps's chief engineer, General Andrew Humphreys.

With the canal plan all but settled, James Eads entered the debate. Nearly forty years earlier, Eads had been working in St. Louis when huge sandbars rose up between the city's waterfront and the river. Eads watched as a young army engineer carried out an ingenious plan. At a strategic point in the river, the officer constructed a great pier, or jetty. The jetty guided the river's currents directly against the sandbars. The sandbars washed away and the threat to the city disappeared. The young officer's name was Robert E. Lee.

Eads had not forgotten. Since then, he had seen jetties clear out the mouths of European rivers, and he had only contempt for the corps's planned canal. So Eads proposed to build a pair of parallel jetties in the southwest pass of the Mississippi. The jetties, he said, would squeeze the river's spreading currents like the nozzle of a firehose and blast the sandbars into the Gulf. He promised that his channel would achieve a substantially greater depth than the corps was proposing for its canal. And he said it would be wide enough that ocean-going ships could pass each other in and out of the river. He could do the job for just $10 million, he said, and the government wouldn't have to pay until he delivered.

Humphreys and a phalanx of critics lashed back. All through the spring and summer of 1874, Eads and Humphreys battled for support in Congress, lobbying, cajoling, arguing, lecturing. Newspapers across the country covered the fight. Civilian engineers supported Eads, hoping to take the corps of engineers down a peg. Humphreys had the backing of New Orleans's business elite. These men dis-

THE BIRTH OF THE BLUES

By the late nineteenth century, the hard work of levee construction and repair was done by black laborers in temporary levee camps. The contractor's white foreman—invariably called "Mister Charlie" by the black mule skinners and laborers—ruled with a pistol and sometimes a hired black enforcer. The camps rang with high, lonesome, musical cries, some with words and rhythm and some without, always with melancholy bent notes and pain-filled moans. According to the musicologist Alan Lomax, these "levee hollers" gave birth to the blues. This was one of them:

"Mister Cholly, Mister Cholly,
Did the money come?"
He say, "The river too foggy,
And the boat don't run,
Oh-oh-oh-oh, the boat don't run."

I ax Mister Cholly just to gimme one dime,
Just gimme one dime.
He say, "Go long, nigger,
You a dime behin,
Oh-oh-oh-oh, you a dime behind."

"Mister Cholly, Mister Cholly,
Just gimme my time."
He say, "Go on, ol nigger,
You time behin,
Oh-oh-oh-oh, you time behind."

trusted "strangers" like Eads, "who can know nothing of our inexorable enemy," the river, which would teach such upstarts "modesty and humility in the presence of the gigantic torrent."

"I am sure I have not learned 'modesty and humility in the presence of the gigantic torrent,'" retorted Eads, who had spent far more time in the Mississippi—literally—than any New Orleans tycoon. "Nor do I believe that it can be controlled by modesty and humility."

After more months of study, a committee of experts appointed by Congress recommended jetties by a vote of six to one. But General Humphreys devised a trap. His friends in Congress insisted that Eads switch his plan to the shallower South Pass—this would be cheaper but more difficult—and when Eads argued, they said he was trying to extract excessive profits from the deal. Eads was forced to accept the South Pass plan.

Defiant, he promised victory.

"If the profession of an engineer were not based upon exact science, I might tremble for the result," he said. "But every atom that moves onward in the river, from the moment it leaves its home amid crystal springs or mountain snows . . . until it is finally lost in the vast waters of the gulf, is controlled by laws as fixed and certain as those which direct the heavenly bodies. . . . I therefore undertake the work based upon the constant ordinances of God Himself; and [if] he will spare my life and faculties for two years more, I will give the Mississippi River . . . a deep, open, safe and permanent outlet to the sea."

In May 1875, Eads set off from New Orleans with a fleet of barges, tugboats, steam launches, and floating pile drivers, all bound for the low-lying landscape where little but whispering grasses and fishermen's huts marked the presence of land in a wilderness of water and sky. They found the sand-clogged South Pass only eight feet deep.

First Eads's men built a settlement of white cottages and plank walks, which they called Port Eads. Then they drove two lines of piles—log posts like telephone poles—deep into the river bottom. These parallel lines of piles stretched about two miles long, about a thousand feet apart. They marked the lines of the two jetties.

Eads had dropped off some of his crews forty miles upriver, where they waded among water moccasins and leeches to cut thousands of young willows. They sent the trees by barge to Port Eads.

Eads's laborers and engineers with their brush mattresses, the foundation of the jetties.

There other crews heaped the trees in pine frames to make giant, brushy "mattresses," each 100 feet long, 35 to 60 feet wide, and 2 feet thick. Dutch dike-builders had invented these odd contraptions, but Eads's method cut the time it took to build one from two days to two hours. When each mattress was finished, a tugboat towed it to the jetty pilings, where men dumped rocky rubble on it, forcing it to sink. Then another mattress was sunk on top of the first, and so on until the mattresses stood up to sixteen layers deep.

Section by section, these towers of sunken willow mattresses formed the jetties' walls—or, more accurately, they formed a dense scaffolding for the walls. The river completed the job, just as Eads had foreseen. Gradually the passing currents deposited sediment on and among the mattresses, filling in the jetties until they were as solid as concrete.

But before he could prove the jetties were deepening the channel, Eads ran short of funds. Humphreys's men on the scene conducted their own depth soundings and claimed the channel was only 12 feet deep—far short of the 30-foot requirement—and that the jetties had created "the nucleus of a new bar" farther out in the Gulf.

Eads's own soundings showed the channel was at least 16 feet deep already, and that the sandbar was a mirage. But federal officials refused to issue the official report of soundings, or to order new ones. Without this official proof Eads would be unable to attract new investors, and his whole endeavor would collapse.

Then, on May 12, 1876, the big oceangoing steamer *Hudson,* with a draft of fourteen feet, seven inches, approached the mouth of the river from the Gulf. Its captain was E. V. Gager, a friend of Eads and a supporter of the jetty plan, who had said he hoped his ship would be first through the new channel. Learning what was at stake, Gager, despite a falling tide, ordered his vessel onward.

James Eads had told him the channel was sixteen feet deep. So Captain Gager ordered his engines put to full steam. Pushing a half-crown of white foam before her, the *Hudson* entered the channel . . . kept moving . . . steamed through.

"Captain Gager . . . greatly assisted the enterprise in one of its darkest hours," Eads's chief assistant wrote later, "for the stubborn facts brought out by his brave action could not be gainsaid."

Investors returned and the project was saved. By that fall the channel's depth reached 20 feet. In 1879 it was 30 feet. Soundings found no new sandbar rising in the Gulf of Mexico. In the five years after Eads began, the tonnage of cargo moving from St. Louis through New Orleans to Europe increased sixty-six fold.

"Human patience and courage and industry, backed by an indomitable and untiring will, and informed and directed by human skill, have applied the force of nature to the accomplishment of an end too vast for mere artificial agencies," a New Orleans editorialist wrote. "Man has used the tremendous river which uncontrolled has been its own oppressor and imprisoner, and has now become its own liberator and saviour."

THE WATER RISES

Eads had made the Mississippi do his bidding with a work of engineering that became famous around the world. That was good and bad. For some people now lost a little of that "humility in the presence of the gigantic torrent" that the Mississippi demanded. They believed Eads's jetties confirmed the levees-only theory of flood control.

Eads himself believed nothing of the kind. He said levees would never cause an appreciable deepening of the riverbed because they

A farmer seeks refuge on a levee with his livestock during an early-twentieth-century flood on the Mississippi.

stood back from the banks, and confined the river only in flood season. Jetties, by contrast, were built right in the river. They didn't merely hold the current off the land. They concentrated it, increasing its power. And they performed this feat year round. Nonetheless, people now thought levees, like jetties, would force the river to scour a deeper channel, which in turn would mean fewer floods.

Levees-only offered a political advantage, too. In the 1870s and 1880s, the levee system lay in ruins. In 1865 the corps of engineers had counted fifty-nine breaks in the levees in the Mississippi Delta alone. Demolition teams under General Ulysses S. Grant had done a good deal of the destruction. The river's natural powers of erosion, plus great floods in 1867 and 1874, did more. And the Civil War had impoverished the states of the lower Mississippi. Few levee districts could afford to do their work well.

worse than 1927. All new works would be judged by their ability to protect the valley from "Project Flood."

At Birds Point, Missouri, where the Ohio tumbles into the Mississippi, the corps built a floodway five miles wide and sixty-five miles long to bypass a chokehold where there isn't enough riverbed to bear the combined flow of the Mississippi and Ohio in flood. A low, "fuse-plug" levee was built as a gate to the floodway. Only a major flood would blow out the fuse plug. Water would then flow into the floodway, run south, and rejoin the main current where the channel widens near New Madrid.

The Bonnet Carré spillway is a key element of the Jadwin plan for flood control. The Corps controls its 350 bays (at right), through which water can be released during a flood. The baffles (center) and articulated slabs (left) slow down the water during floods.

If floodways had been Ellet's pet idea, the corps borrowed other theories from James Eads. They built dikes at key points to stabilize the current. Between the mouth of the Arkansas and the mouth of the Red, they dug a straight path through a tortuous series of horseshoe bends, shortening the river by some 150 miles. (The longer the river, the more friction the water encounters along the banks. Friction slows the current and forces the river to rise higher on the levees.) These "cutoffs" were so successful they eliminated the need for an entire additional floodway.

Thirty miles above New Orleans, at a place called Bonnet Carré, the Mississippi had blown crevasses in the east bank levee again and again, streaming across seven miles of field and semitropical swamp to reach Lake Pontchartrain, a natural outlet to the Gulf of Mexico. Since 1816 people had been saying it made sense to let the river have its way here. The 1927 flood demolished opposition to the idea, and a spillway at Bonnet Carré was incorporated in the Jadwin Plan.

Using a giant laboratory model to test their plans, corps engineers designed a spillway that closely resembled an irrigation dam. This would not be a fuse-plug structure like Birds Point, designed to give way under the pressure of high water. Instead, engineers would carefully control the water flowing through it. Set back from the river, the spillway would run parallel to the riverbank for a mile and a half, with some 350 gates made of rough-hewn timbers called "needles." When a dangerous flood crest approached, cranes could pull needles up and out of the spillway—as many or as few as necessary. These gates, once open, would skim the crest off the river.

But that carried its own dangers. Floodwaters would plunge

through the gates with monumental force, gouging a canyon in the earth. So engineers found ways to calm the water down. On the side of the spillway away from the river, they laid a broad, low apron of concrete called a stilling basin. Here the water would crash into a big belt of triangular baffles that looked like waist-high dragon's teeth. These would dissipate the water's energy without blocking the flow. Beyond the baffles, the water would cascade onto a concrete mat that resembled a giant waffle. Its channels would further subdue the flow, and the water, robbed of its power to destroy, would run more or less placidly down the seven-mile slope to Lake Pontchartrain.

By early 1931 constructions teams finished the spillway. It was a structure to overwhelm the eyes; one could see the whole thing only from the air. Then they built bridges to carry two rail lines and a highway over the long floodway to the lake. Finally they raised a fuse-plug levee at the edge of the Mississippi, across the spillway's front, and covered it with Bermuda grass sod.

In December 1936 the entire structure was declared complete. By Christmas it was raining in torrents up and down the Mississippi valley. The annual high waters had spared the valley for ten years. But now the sod on the spillway levee would not take root before Bonnet Carré, and the rest of the Jadwin Plan, met its first test.

Soon a flow of 1.85 million cubic feet per second was recorded on the Ohio at Cairo, a spectacular high water. The engineers of the corps watched intently as the water surged against the new fuse-plug levee at Birds Point. When it failed to break, they dynamited it. That saved Cairo. As the crest moved south it caused small floods in backwater regions, but the strengthened levees on the main river held.

At Natchez, Mississippi, two hundred miles north of New Orleans, the flood gauge recorded its highest level ever. The same thing happened at Carrollton. That was the official trip-hammer for Bonnet Carré. The corps gave the word.

On February 18, 1937, with crowds watching, the cranes atop the spillway hoisted timber needles out of 285 of the 350 bays. Water rushed into the stilling basin and exploded through the baffles. Surging down the floodway toward the lake, the tide wrenched trees from the soil. Yet even that degree of force lay within the margin of safety. For one week, then two, the Carrollton gauge stayed steady, and New Orleans stayed dry. By March 7 the river was falling, and the corps began to reinsert the needles. Nine days later the spillway was closed.

The Bonnet Carré spillway decreases the threat of uncontrolled levee breaks like this one.

The 1937 high water was not the Project Flood. But it was high enough to prove Bonnet Carré a signal success, and to show that the Mississippi might tolerate man's effort at a compromise. Not a single levee on the lower Mississippi failed in the 1937 flood, nor has any failed since.

A PATIENT RIVAL

If the Mississippi is a toddler in geological age, the Atchafalaya (uh-CHAFF-uh-LIE-uh), which connects to the Mississippi via a manmade waterway some 120 air miles northwest of New Orleans, is a newborn infant. It emerged from the swamp in the 1400s, carrying water from a horseshoe bend in the Mississippi (later called Turnbull's Bend) through a jungle labyrinth to the Gulf. To complicate the landscape, the Red River emptied into Turnbull's Bend a few miles north of the Atchafalaya. To help steamboat traffic, Henry Shreve dredged a cut-off canal through Turnbull's Bend in 1831. But one seven-mile arm of the bend—soon called Old River by the locals—remained to connect the Mississippi to the Atchafalaya.

The Atchafalaya is what geologists called a "distributary" river, meaning it carries water *out* of a larger stream. But a distributary can become a trickster and a thief. As the main stream rises atop its floor of silt, it sends more and more runoff spilling down the distributary. That runoff water gradually digs the distributary a deeper bed. In time, the distributary offers a deeper, steeper outlet to the sea. One big flood and the main stream may change its mind entirely. Suddenly the old channel is abandoned and the river is following the distributary's path to the ocean.

Over the eons this has happened repeatedly at the Mississippi's mouth. At least five now quiet bayous have all served their time as the main channel. The last shift occurred about A.D. 1000. If the Atchafalaya were to capture the Mississippi now, the Port of New Orleans, one of the world's busiest, and a multibillion-dollar belt of riverfront industry would be stranded on a brook.

River engineers saw this coming. Just after the 1927 flood,

Colonel Charles Potter, president of the Mississippi River Commission, looked over the big river and its eager distributary and said the Mississippi was "just itching to go that way."

But the corps of engineers could hardly wall off the Atchafalaya, which was by far the most important outlet envisioned in the Jadwin Plan, much bigger even than Bonnet Carré. Plug the Atchafalaya and you not only devastate the fisheries and Cajun communities that depend on it for water, you also doom New Orleans in the next great flood. Yet every gallon of water spilling into the Atchafalaya dug its bed a little deeper and raised the likelihood of an eventual shift. By 1950, geologists said the slope of the Atchafalaya's path to the Gulf

The low sill and overbank structures at Old River, under construction in 1959, were designed to regulate the Mississippi's flow into the Atchafalaya River.

had become three times steeper than the Mississippi's current channel. The river, they said, would certainly shift from one to the other by 1975.

So the corps constructed another compromise. Earthmovers dug a new, seven-mile-long channel—a new Old River—between the Mississippi and the Atchafalaya. Across this channel the corps built a 566-foot version of the Bonnet Carré spillway. They called it the Old River Control Structure, usually shortened simply to "Old River Control." Engineers raised its gates just high enough to allow 30 percent of the Mississippi to enter the Atchafalaya. That was judged to be enough but not too much. In the Atchafalaya swamps, the corps constructed a system of floodways, chiefly the West Atchafalaya Floodway and, thirty miles to the south, the Morganza Floodway. Then, in 1963, the engineers blocked the old Old River, opened the new one, and watched water cascade through the carefully calibrated gates of Old River Control.

Through the 1950s, '60s, and early '70s, the Mississippi cooperated. The spring high waters were manageable.

Early in 1973, major storms raised the Missouri and the Tennessee to towering levels. The crest approached.

"This is not a routine high water," General Charles Noble, chief engineer of the corps's Lower Mississippi Valley Division, warned. "We are confronted with river conditions which, if not controlled, could cause more loss of life and property than this valley experienced in the 1927 flood."

Corps engineers watched the water pound through Old River Control. On April 8, for only the fourth time in its history, they opened the Bonnet Carré spillway downriver. Yet people standing on the 200,000-ton structure at Old River felt it shake.

As the journalist John McPhee told it, a fisherman came into the office of LeRoy Dugas, who manned the controls at Old River, and asked: "Is that south wall supposed to be moving like that?"

"What do you mean?" Dugas asked.

"On the control structure," the man said, "a wall on the other side is moving."

"No," Dugas replied. "It's not supposed to be moving."

Soon that wall gave way and crumpled into the Mississippi. On the Old River side of the structure, the water seized seven-ton boulders and hurled them downstream. Unseen below the roiling surface, mad currents were digging a hole the size and shape of a football stadium, and scouring cavities under the structure itself.

The corps waited a little while longer, then for the first time opened the gates of the new spillway downriver at Morganza.

Old River Control held—barely. If it had collapsed, engineers believe 70 percent of the Mississippi would have run down the Atchafalaya slope. The struggle to keep the river in its old channel would have been lost.

Humility and Ingenuity

Scientists at Louisiana State University declared in 1980 that a shift in the Mississippi's course remained "simply a matter of time." But the people who lead the U.S. Army Corps of Engineers are professional soldiers. They do what their country tells them to do, whatever the odds, and their country has told them to keep the lower Mississippi where it is. So in the wake of the great scare of 1973, they went back to work at Old River.

Corps workmen drilled holes through the damaged control structure and pumped truckloads of cement grout into the cavities underneath. Into the cavernous craters just beyond the structure, they dumped 185,000 tons of rock. That was enough to hold the

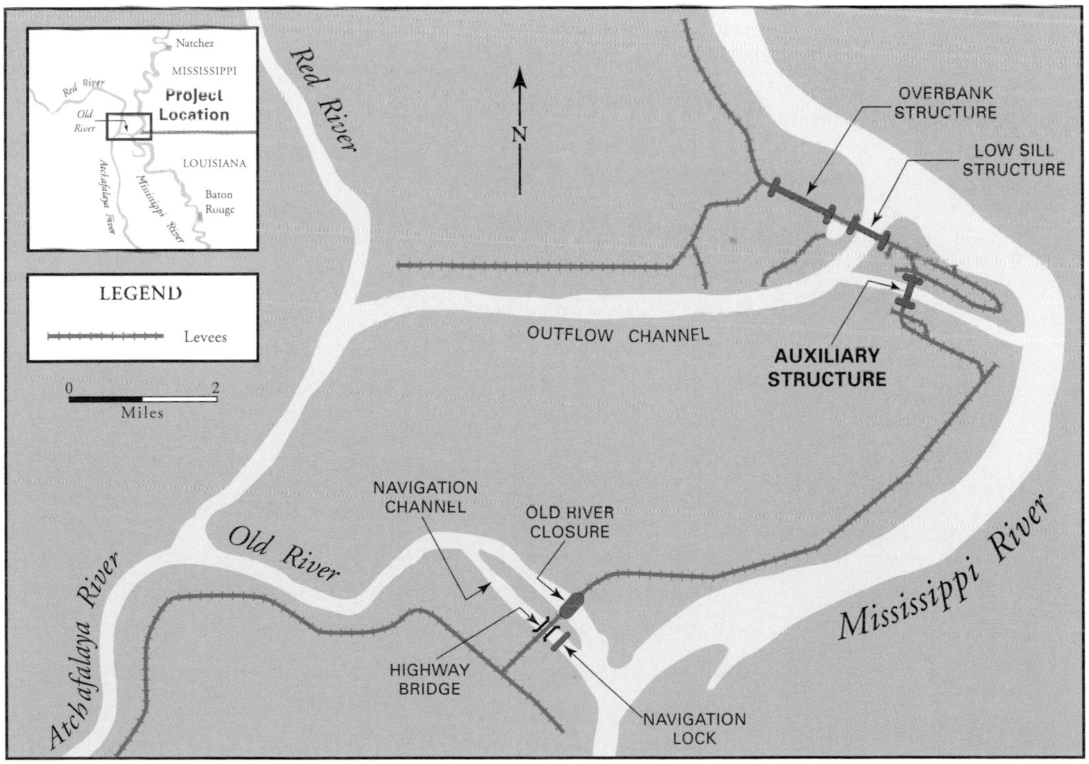

Old River Control.

"AN IMBECILE BLIND TITAN"

William Alexander Percy was the scion of a Mississippi River dynasty. His grandfather settled the Delta. His father, U.S. Senator LeRoy Percy, was the most powerful man in the region from the 1890s until his death in 1929. His cousin and adopted son, Walker Percy, would join the highest echelon of American novelists.

War hero, poet, lawyer, and planter, Will Percy gave his own gift to American letters: a memoir entitled Lanterns on the Levee, *first published in 1941. Gracefully written, painfully evocative of a lost way of life, often racist by later standards, it includes this description of the 1927 flood from the vantage point of Greenville, Mississippi, in the heart of the flood zone:*

[D]uring every high-water scare Delta citizens walk the levee all night with pistol and lantern, nowadays with flash-light. . . . During these times the river is a savage clawing thing, right at the top of the levee and sounding at night like the swish of a sword or the snarl of a beast. It puts ice in your heart. . . . Each guard walks alone, and the tiny halo of his lantern makes our fearful hearts stouter. . . .

The greatest flood in the history of the Mississippi was roaring south between levees that trembled when you walked on them. The workers knew the fight was well-nigh hopeless, but there was nothing else to do but fight. . . . In the glare of improvised flares and flood-lights they swarmed over the weak spot like ants over an invaded anthill. But about daylight, while the distraught engineers and labor bosses hurried and consulted and bawled commands, while the five thousand Negroes with . . . hundred-pound sandbags on their shoulders trotted in long converging lines to the threatened point, the river pushed, and the great dike dissolved under their feet. The terrible wall of water like an imbecile blind Titan strode triumphantly into our country. . . . If the Lord was trying to cement us with disaster, He used a heavy trowel that night.

The 1927 flood was a torrent ten feet deep the size of Rhode Island; it was thirty-six hours coming and four months going. . . . Between the torrent and the river ran the levee, dry on the land side and on the top. The south Delta became seventy-five hundred square miles of mill-race in which one hundred and twenty thousand human beings and one hundred thousand animals squirmed and bobbed.

In the thirty-six hours which the river required . . . to submerge the country, panicky people poured out of Greenville by the last trains and by automobiles over roads axle-deep in water. These were mostly frantic mothers with their children, non-residents from the hills who regarded the river hysterically and not devotedly, and the usual run of rabbit folk who absent themselves in every emergency. During the same hours of grace panicky people poured into Greenville. These were mostly Negroes in dilapidated Fords, on the running-boards of trucks, or afoot carrying babies, leading children, and pulling cows, who are always at their worst in crises. . . .

For thirty-six hours the Delta was covered in turmoil, in movement, in terror. Then the waters covered everything, the turmoil ceased, and a great quiet settled down; the stock which had not reached the levee had been drowned; the owners of second-story houses with their pantries and kitchens had moved upstairs; those in one-story houses had taken to the roofs and the trees. Over everything was silence, deadlier because of the strange cold sound of the currents gnawing at foundations, hissing against walls, creaming and clawing over obstructions.

When at midnight the siren of the fire department by a long maniac scream announced to the sleepless town that the water had crowded over the town's own small protection levee, we knew the last haven of refuge had been lost. In each home the haggard family did its hysterical best to save itself and to provide for the morrow. Outside on the sidewalks you heard people running, not crying out or calling to one another, but running madly and silently to get to safety or to their loved ones. It was a sound that made you want to cry.

structure in place, even when another major flood rolled through in 1983. A few hundred yards away, the corps built an Old River Auxiliary Control Structure. Costing nearly as much as the original expenditure for the entire Jadwin Plan, it comprised 7 towers and 6 gates, each 62 feet wide.

"I was asked, when I was testifying before Congress about the auxiliary structure at Old River, if this improvement would be all that was ever needed there," recalled Fred Bayley, former director of the corps's engineering division for the Lower Mississippi Valley District. "And my reply was, 'No, sir.' And the next question was, 'Well, what will be the next thing?' I said, 'I don't know, but with the system we're living in out here, there will be something, and that you can count on and that you must ever be ready for.'"

Corps scientists sought new approaches to their great problem. They turned their attention to a phenomenon that James Eads had discovered firsthand more than a century earlier, under a diving bell—the sand that "drifted like a dense snowstorm at the bottom." Sediment, after all, was the root of the problem. There was too much in the Mississippi and not enough in the Atchafalaya. If a lot of it could be moved from the former to the latter, the fatal percentages might shift back in the Mississippi's favor.

So a continuous dredging program was undertaken at Old River. Sand is scooped from the Mississippi's silt load and dumped into the channel that flows to the Atchafalaya. No one knows whether it will be enough. "Mother Nature doesn't holler back at you that quick," says a corps hydrologist. When that message comes, the corps will be listening closely. They know, as James Eads said, that "modesty and humility" will not control the river. But they have learned that a little humility is wise where the Mississippi is concerned.

When people came to live by the Mississippi, the struggle to control it became inevitable. Only nomads could live there if the river ran free, and Americans did not intend to be nomads. It may be true, as a historian of the corps of engineers once observed, that in the Mississippi valley, "nature might reasonably have asked a few more eons to finish a work of creation that was incomplete." But Americans are impatient. They would just as soon finish Creation on their own.

CHAPTER TWO

THE COLORADO

John Wesley Powell exploring near the Grand Canyon with Chief Tau-Gu of the Paiute Tribe in 1873. The tribe called Powell Ka-pur-ats, meaning "one arm off."

Hoover Dam, . . . the several million tons of concrete that made the Southwest plausible, the fait accompli that was to convey, in the innocent time of its construction, the notion that mankind's brightest promise lay in American engineering.

— *Joan Didion, "At the Dam"*

Born in 1834, John Wesley Powell grew up, like many of his generation, looking west, but not for gold or fresh farmland. He was infected with a scientist's yearning to probe the natural secrets of his half-explored country—a passion that would raise him to preeminence among American naturalists and make him perhaps the one true prophet of the American West.

Raised on a succession of frontier farms in the Midwest, Powell taught himself more natural history by the time he was twenty-five than most Harvard scholars ever learned. The Confederate ball that took off his right arm at Shiloh did nothing to dampen his zeal for science, and in 1867, now a prairie professor without a college degree, he recruited a few volunteers and set off for the El Dorado of American naturalists, the trans-Mississippi West.

For more than half a century, American soldiers, miners, trappers, cattle drivers, railroad surveyors, and settlers had crisscrossed the West. Yet it remained largely unmapped and poorly understood. Americans knew least of all about the great southwestern region they had won from Mexico in 1848—only that it was hot and dry, and that a river ran through it, a strange, narrow, twisting river that cut unfathomed gorges through plateaus of stone. By the name a Spanish priest had scrawled on a map a century earlier, they knew the river ran red: Rio Colorado. But no one of European heritage had traveled its entire length.

In this unexplored region, Powell's countrymen in the East hoped to plant a garden. After all, the desire to make a new society of pastoral bounty and freedom had passed from one generation to the next since the Puritans—to Thomas Jefferson, who bought an

"empire for liberty" from France; to the plowmen who cleared Appalachian valleys and midwestern forests; to Brigham Young and his Mormons, fleeing persecution; to the congressmen who demanded the seizure of California and the Southwest in the name of "manifest destiny" and promised western homesteads to settlers.

The argument over how the new lands should be cultivated—by slaves or free men—had led to civil war. With that issue decided, and untilled land all but gone in the East, Americans looked for a new fresh-start frontier. A cyclical increase in western rainfall, beginning in the mid-1860s, nourished their hopes. In fact, wishful thinking became so strong—with loud encouragement from railroad promoters and land speculators—that people began to believe western emigration was actually coaxing more moisture from the skies. A quack scientist's assertion that "rain follows the plow" became an article of faith. Settlers crossed the Missouri River in hordes, embracing a vision of dry fields that soon would "blossom like the rose."

John Wesley Powell now sought to understand what this myth-shrouded West was really like. His first two trips to the Rockies, in 1867 and 1868, gained him attention and backing for

The artist and topographer on Lt. Joseph Ives's early expedition on the Colorado, Baron F. W. von Egloffstein, made this etching of Black Canyon, just above the eventual site of Hoover Dam, for Ives's *1861 Report upon the Colorado River of the West to the Secretary of War.* The height of the canyon was exaggerated, perhaps understandably.

John Wesley Powell, seated in the middle of the boat to the far left, leads a scouting expedition in the Colorado region.

a far more ambitious expedition. With a small crew of semiprepared recruits, he made a one-hundred-day trip by boat from the headwaters of the Green River in Wyoming through the terrifying rapids of the Colorado all the way to Grand Wash Cliffs near the Utah-Nevada line. They were the first to survey the river's fabulous canyons and weird landscapes—"a wilderness of rocks, deep gorges where the rivers are lost below cliffs and towers and pinnacles and ten thousand strangely carved forms in every direction."

Acclaimed as a hero, Powell became head of the U.S. Geological and Geographical Survey of the Rocky Mountains. For the next decade he drew maps, measured rainfall, judged elevations, examined plants, chipped fragments of rock, studied native tribes (his particular passion), and visited Mormon villages where farmers raised crops with creative, communal techniques. By 1878 Powell knew more about the West than anyone else, and he delivered his verdict to Congress in a two-hundred-page broadside bearing the soporific title "Report on the Lands of the Arid Regions of the United States."

The West was beautiful, he said. But it was no garden.

Powell declared that the hundredth meridian—the north-south line piercing the nation from the Dakotas to central Texas—was the border between two starkly different worlds. To the east was the original America, a land of prosperous farmers enjoying all the rain they needed. To the west, most land was simply too dry for conventional agriculture. "Rain follows the plow" was a crock. Yet under the federal system of land distribution, the West was being settled as if it were the East, block by rectangular block, regardless of access to rivers. Under existing law, whoever owned land upstream on a western watercourse could take the water and flourish while his downstream neighbors wilted in the sun. And anyone buying land far from a watercourse stood no chance at all. In the West, Powell said, the standard American homestead of 160 acres would kill a solitary farmer before it would give him a dependable living. Washington was courting disaster, both personal and civil.

Powell believed as deeply as anyone in the Jeffersonian ideal, the notion that independent yeoman farmers were the mainstay of the

Republic. And *some* of the West could be farmed, he said; it wasn't barren, just dry. Careful planning and irrigation could make those lands sprout vegetables, fruit, and grain. But the lands must be managed with the communal spirit of New England barn raisings and the ingenuity of Mormon irrigationists.

To this end, Powell proposed a revolution in federal land and water policy. Most western land should be designated as suitable only for mining, logging, and open-range livestock. Land near the rivers, the only real farmland, must be divided into small tracts, each with river access to permit irrigation. The all-important power over water rights and irrigation should be vested neither in government nor in private corporations, but in cooperative associations of neighbors. In short, westerners must enshrine the conservation of water as their central social principle. Wherever water was at stake, the community must take precedence over the individual.

Fifty years would go by before one of mankind's masterworks rose up to fulfill Powell's vision. In his own day, his countrymen saluted his bravery, thanked him for his maps, and ignored his advice. They went on settling the West in the old patterns.

Then, in the 1880s, the abnormal cycle of wet years ended, and farm families straggled back toward the East, fleeing hot winds, dust storms, and depression. From Montana to Mexico, "man has retired before hostile nature," the historian Edwin Erle Sparks observed from his vantage point at the University of Chicago. "Abandoned dugouts or sod houses show where over-confident man has retreated from the unequal contest."

This great defeat failed to stifle the westering impulse. But it did force Americans to approach the West with new tools. In the 1890s, they founded a movement that was part politics, part technology—the reclamation movement. A century later, its original zeal is forgotten, but for a long time the reclamation idea wielded enormous clout from the Pacific coast all the way to Washington. Its very name suggests Americans' proprietary approach to the West; the idea was not to *claim* the arid lands for cultivation and settlement, but to *reclaim* them, as if a treacherous Nature had stolen them from their rightful owner, the American plowman. Irrigation on a heroic scale would accomplish the miracle. Engineers and farmers would fight shoulder to shoulder against the desert. The old dream hadn't died at all. If the West wasn't a garden, if rain didn't follow the plow, then Americans would *make* a garden there.

Look to the rivers, John Wesley Powell had said—there lay the West's true riches, in the Missouri, the Columbia, the Platte, the Rio Grande, and especially his own mighty Colorado. Turning the water on the arid lands was a monumental challenge. Whoever mastered it would make history. Or at least a fortune. Those who glimpsed the fortune got to the water first.

THE FATE OF SMALL PLANS

Modern industries are handling the forces of nature on a stupendous scale . . . the lightnings are harnessed and floods are tamed. Woe to the people who trust these powers to the hands of fools!

—John Wesley Powell, 1889

The Colorado, like the Nile, is what geologists call an "exotic" river. Such streams trace their sources to a place of heavy rain or snow (in this case the Continental Divide in Wyoming); the water streams down the mountains, then flows across a desert to the sea. Like other exotics, the Colorado lurches between drought and flood. When mountain snows melt, the river rises spectacularly and speeds toward the Pacific in torrents. Then it dwindles to a comparative trickle for much of the rest of the year.

Charles Rockwood was the first engineer to irrigate southern California with Colorado River water. The result was disaster.

Like the Mississippi, the Colorado carries silt downstream—stony red silt that helped the river cut canyons in the sandstone and shale that rose in its path. Also like the Mississippi, the river wandered back and forth near its mouth at the Gulf of California. As it wandered, it deposited a wide ridge of silt, and the river ran along the top of that ridge. In flood, the river threatened to spill down the ridge's western slope. Scientists of the 1800s spoke of its "whims" and "viciousness." Anyone who tried to control it would meet a strong and unpredictable adversary.

In 1892, the first great effort to do so started in the mind of an ambitious irrigation engineer named Charles Robinson Rockwood. Burly and bulldog-faced, Rockwood was scouting the region where California, Arizona, and Mexico meet—the Colorado delta. He looked westward, down the gentle alluvial slope that ran for miles into southeastern California. Here, Rockwood realized, was a landscape that might have been designed for irrigation on an epic scale.

Rockwood saw what a sprinkling of Spanish and American travelers had seen for three centuries: 1,600 square miles of uninhabited sand in the shape of a broad, shallow basin. Over millions of years, the course of the Colorado periodically had shifted into this basin, then abandoned it, leaving a pool of water nearly as big as the Great Salt Lake to evaporate in the sun. Americans called the basin—which was empty of water at this moment in geological history—the Salton Sink. The Spaniards used a more hopeful name: *La Palma de la Mano de Dios,* the Palm of the Hand of God.

God hardly seemed generous here. It was about the sunniest, hottest place in the United States. A gold prospector who crossed it in the rush of 1849–50 called it "scorching and sterile—a country of burning salt plains and shifting hills of sand, where the only signs of human habitation were the bones of animals and men scattered along the trails." More rain fell on some parts of the planet in twelve months than fell here in seventy-five years.

Yet Rockwood perceived an obvious geological fact—that the region lay below the elevation of the Colorado River—and the import of that fact. If water could be coaxed away from the river's bed, it would flow downhill to the Palm of the Hand of God, which might then support crops all the year round.

After a seven-year struggle for financial backing, Rockwood formed the California Development Company (CDC). Through a tangled string of shady dealings, he and his allies seized the legal rights to the water of the Colorado and began to squeeze a fortune out of them. With a few taps on a typewriter, the bleak-sounding "Salton Sink" became the "Imperial Valley." Leaflets invited settlers to an agricultural haven of cheap land and water. Buyers arrived by company-chartered trains and creaking wooden wagons, some of them eager for an irrigated homestead, some just eager to buy low and sell high once irrigation had turned desert to garden.

With artful promises, early promoters recast the Salton Sink as an agricultural paradise.

On Thanksgiving Day 1900, starting at a point on the western bank of the Colorado across from Yuma, Arizona, CDC crews began to dredge a four-mile canal straight south into Mexico. Just over the border, in Mexican territory, the water would flow into a dry channel called the Alamo Barranca. Long ago, the Colorado had followed this route into the Salton Sink. Now, once again, the old Alamo channel would carry the water west for some forty miles, where more waterways, natural and manmade, would carry it to new farms in the Imperial Valley. Some four hundred miles of irrigation ditches were dug to water the farms. Back at the point where Rockwood's manmade canal diverged from the Colorado, workmen erected wooden headgates. These could be opened and closed to control the flow.

Just six months later, on May 14, 1901, freshets of Colorado

One of the main canals dug by Rockwood's California Development Company in the Imperial Valley.

The shrubbery near new houses in the Imperial Valley proved the desert was in bloom.

River water came tumbling into the powder-dry Salton Sink. The farmers watched in astonishment as the endless sun and bountiful water nourished fast-maturing grapes, melons, and more than one bale per acre of prized long-staple Egyptian cotton. In the first year alone they cut six crops of alfalfa. Stock in the CDC soared. More settlers came in a swarm. When E. H. Harriman, owner of the Southern Pacific Railroad, laid a line into the valley, a great future seemed assured. By 1904, farmers were cultivating seventy-five thousand acres.

Then the Colorado reminded everyone how it got its name. What its red silt had given, it now began to take away. Every year, some 160 million tons of sediment flowed into the Colorado delta. When Rockwood diverted water into his canal, some of that sediment came with it and settled on the bottom, grain upon grain. After three years, the first four miles of the canal were a highway of sludge. Imperial Valley crops began to die of thirst, and farmers began to sue for damages. At the same time, agents of the federal government were challenging the CDC's right to appropriate so much water. "These gentlemen," said an opponent of the CDC, "have . . . claimed the melting snows of the Rocky Mountains as their property."

What followed was a tragicomedy of folly and persistence. Rockwood had not only ignored the silt in the river. He had chosen to fool with the river's course at the very point in geological time when conditions were ripe for the river to change course on its own. Far from controlling the river, Rockwood was helping it along in the direction of disaster.

Mindful for the moment of only his muddy clog, Rockwood crossed the border into Mexico, where the feds couldn't bother the CDC. Without clear Mexican authority, he dug a new canal, just a quarter-mile long through soft sand. This ditch bypassed the clog in Rockwood's old canal. Once again, water flowed to the farmers. But for whatever reason—red tape with the Mexican authorities or a shortage of money—Rockwood failed to install a control gate across

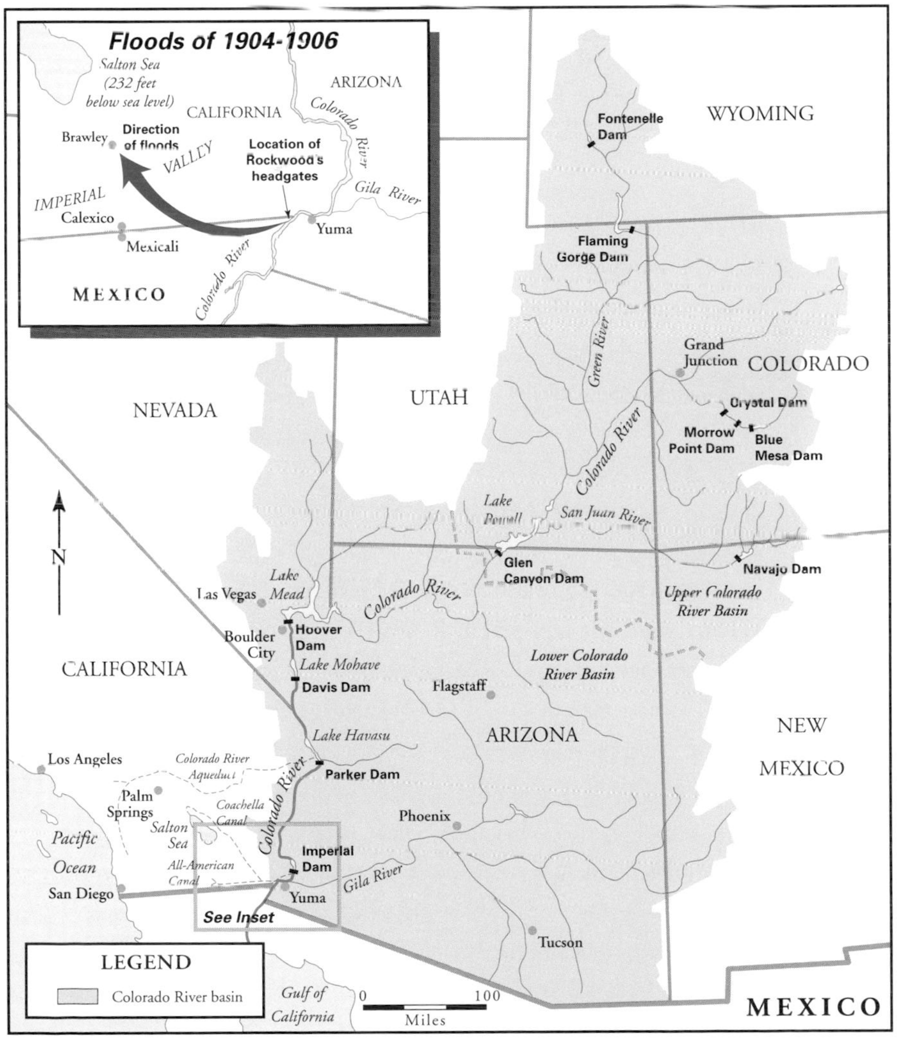

The Colorado River Basin.

A wooden headgate installed by Rockwood to regulate the flow of the Colorado River into canals. Such structures were no match for the river in flood.

The devastation wrought by the floods of 1905.

this new opening in the Colorado's bank. He had no means for regulating the flow. He thought he could install a headgate before the next spring floods.

The Colorado didn't wait that long. In all the previous twenty-seven years, the river had flooded in winter only three times. In the late winter of 1905 alone, it flooded five times. In his own defense, Rockwood later said: "I doubt whether anyone should be accused of negligence . . . in failing to foresee that which had never happened before."

In February 1905, two flood crests swirled through the gateless Mexican canal intake and rushed down the irrigation network to the Imperial Valley. For a time the farmers hurrahed the welcome water. When a third crest arrived in March, it widened Rockwood's new intake to sixty feet. This third flood, Rockwood said, "was sufficient notice to us that we were up against a very unusual season." He directed his work crews to hurl brush and sandbags into the intake to make a hasty dam. But a fourth flood crest shoved the dam away. They tried again; the dam fell before a fifth high water. The intake was now 160 feet wide.

E. H. Harriman, president of the Southern Pacific Railroad, simultaneously faced earthquake in San Francisco and flood in southern California's Imperial Valley.

By summer 1905, most of the Colorado River was flowing into the Imperial Valley, lapping at towns and green fields. At the deepest part of the Salton Sink basin, a new Salton Sea appeared—a vast inland lake of muddy water with no outlet to the ocean.

Grasping for some way out of his predicament, Rockwood approached E. H. Harriman, whose rail line in the Imperial Valley would soon be underwater. Jonah may have approached the whale with more confidence. In 1905, Harriman was becoming to railroads what J. P. Morgan was to banking and John D. Rockefeller was to oil. A superb organizer, ruthless and cold, with a natural inclination to control whatever he touched, Harriman listened as Rockwood described the situation in the valley and asked for a loan of two hundred thousand dollars, arguing the money would safeguard Harriman's interests. Harriman agreed to advance the money in exchange for a controlling interest in the CDC, to be sure the loan was repaid. He installed his own man, Epes Randolph, as president of the CDC, but retained Rockwood as project engineer.

Epes Randolph went to look for himself, taking along his own engineer, Harry Thomas Cory, former dean of engineering at the University of Cincinnati. They could hardly believe what they saw. Cory concluded that Rockwood's decision to dig the Mexican intake

Erosion in the form of rapid cutbacks caused massive destruction.

without installing a control gate "was a blunder so serious as to be practically criminal." Rockwood tried to reassure him, saying the Imperial Valley farms were in no real danger. Cory told Rockwood he'd been playing with fire and set to work to put out the blaze, not yet realizing he had met the challenge of his life.

Cory and Rockwood spent October and November of 1905 building a six-hundred-foot dam across the Mexican intake, which was now, in fact, a crevasse—a gaping hole torn by the river itself. As they finished the dam, an enormous flood crest came sweeping down the Gila River, which joins the Colorado just above Rockwood's canal. In a matter of hours, the Colorado's flow rose from 12,000 to 115,000 cubic feet of water per second. The dam crumbled, leaving the Mexican crevasse as wide as two football fields. The Salton Sea, already spreading over 150 square miles, washed over farms and covered the tracks of the Southern Pacific line.

Over the winter, the engineers drafted a more ambitious plan. They dared not contain the Colorado entirely to its old riverbed. That would turn the Imperial Valley back into a desert, depriving its new farmlands of moisture and its new residents of drinking water. Irrigation was still needed; it just had to be better managed. So they planned a dual assault. They proposed to bring an 850-ton floating steam dredge into Rockwood's silt-packed original canal (the one with its intake in American territory); this monster would dig out the silt clog and restore that path for the water to flow to the Imperial Valley. A strong headgate of steel and concrete—the sort that Rockwood should have built at the outset—would be installed to control the flow. This flow, in turn, would lower the level of water surging through the Mexican crevasse, making it easier to close the crevasse with a dam or levee.

Then, on April 18, 1906, while spring floodwaters were rising on the upper Colorado, earthquake and fire all but destroyed the city of San Francisco. To do the work of rescue and rebuilding, the resources of Harriman's railroad network would be taxed to their limits. Yet far to the south, more than the Imperial Valley was now in peril. In the channel where the Colorado was pounding into the Salton Sea, a little waterfall had formed. In soil as fine as powdered sugar, the lip of the falls collapsed under the heavy flow, then collapsed again and again—a "cutback" that stumbled backward through the desert at a speed of better than a mile a day, deepening and spreading as it went. It was "like a ripsaw tearing through rotten

The Southern Pacific sent carloads of rock to put the river back in its bed.

wood," one historian said. If the cutback reached the Mexican crevasse, the falls there would stand three hundred feet high. The plunging power of the river would scour a trough so deep that no human agency could restore it to its old channel. Some even thought the cutback might continue beyond the intake, carving a new Grand Canyon and destroying large sections of western Arizona and southeastern California. In any case, geologic changes that normally would require centuries were happening in weeks.

In the smoking ruins of San Francisco, Harriman learned of this new threat and advanced another quarter-million dollars to tame the Colorado. Charles Rockwood, at last, was cast out in disgrace, and H. T. Cory assumed full responsibility as chief engineer. Cory now faced a crevasse that was half a mile wide. "The situation was a desperate one and without engineering parallel," Cory said later, "and . . . there seemed to be little more than a fighting chance of controlling the river." As Harriman's first biographer put it, "Nobody had ever before tried to control a rush of 360,000,000 cubic feet of water per hour, down a four-hundred-foot slope of easily eroded silt, into a basin big enough to hold Long Island Sound. There was nothing in the past experience of the world that could suggest a practicable method of dealing with such conditions."

During the spring and summer of 1906, Cory and Epes Randolph made new plans. They would try to block the river with rock instead of brush. They laid a railroad line leading directly to the Mexican crevasse. They amassed supplies—300 Union Pacific "battleship" cars with side-dump doors; rock and clay from every quarry within 400 miles; 1,100 90-foot piles; 19,000 feet of timbers; 40 miles of steel cables; pile drivers; steam shovels. They drafted an army of Native Americans, Mexicans, and "drifting adventurers . . . attracted to the place by the novelty of the work and the publicity given to it in the newspapers." Under martial law enforced by Mexican soldiers, the workers sank a giant mattress of cables, baling wire, and brush to the floor of the Mexican crevasse. Then they built a railroad trestle across the mattress. Cory had the "battleships" hauled onto the trestle, where they dumped their sixty-ton burdens of rock into the river to create a makeshift dam. A new headgate was installed. By the end of September, the dam was high enough. When they were done, the river was contained. For a few days water no longer flowed into the Salton Sea.

On October 11, 1906, the river broke through again.

The engineers and their army started over—more rail trestles, more rock, and massive levees to supplement some old CDC levees. For four weeks, the river again was under control. But on December

A train risks the floodwaters to bring more rock to the break.

7, a torrent descended the Gila River and swelled the Colorado. The flood crest let the new dam stand, but it tore through the old CDC levee and struck off for the Salton Sea in a new channel of its own making.

Harriman and his men now asked whose problem this was, anyway. After all, the U.S. government owned most of the land in the area, and the cutback was threatening a U.S. dam under construction north of Yuma. All the Southern Pacific had at stake was $30,000 in annual rail traffic; it could move its line for $60,000 and be done with it. And the S.P. hadn't caused the fiasco; the CDC had. "It does not seem fair," Harriman wrote President Theodore Roosevelt, "that we should be called to do more than join in to help the settlers."

As it happened, Harriman and Roosevelt, former neighbors and friends in New York politics, recently had gone to war with each other over railroad regulation. The Interstate Commerce Commission was prosecuting Harriman, and only a few weeks earlier the president had referred to the railroad baron as an "undesirable citizen." Now he fired a cable back to Harriman: "It seems to me clear that it is the imperative duty of the California Development Company [which Harriman now owned, more or less] to close this break at once. . . . There is not the slightest excuse for the California Development Company waiting an hour for the action of the government." He promised only to ask Congress, when it reconvened, to approve an "equitable distribution of the burden," which soon exceeded $3 million. (T.R. kept that promise, but Congress did nothing.) So Harriman told his engineer to go ahead and do whatever it took.

"It was simply a case of putting rock into that break faster than the river could take it away," Cory said later. There was no time to sink new brush mattresses. Skeptics said dams built without the mattresses for flooring would sink away into the silt, but Cory didn't think so. He constructed two new rail trestles over the break in the levee and then created the proverbial immovable object—80,000 cubic yards of rock. It took fifteen days. "In that time," Cory said, "I suppose we handled rock faster than it was ever handled before." Twelve thousand miles of western railroads were tied up while carloads of rock came to Cory's trestles from up and down the Rocky Mountains.

In a monument to academic understatement, Cory later concluded his account of the fight by reporting, "The water was shut off on February 10, [1907] at 11 p.m. . . . The event showed the existence and nature of the danger and the necessity for guarding against it." Francis Sellew, a fellow engineer who watched Cory's crews at work, was more emphatic: "Turning the Colorado out of Salton Sea was not so much an engineering triumph as a triumph of engineers, of red-blooded fighters, who did not know when they were whipped."

A. P. Davis, nephew of John Wesley Powell, became the director of the Bureau of Reclamation in 1914 and was the first to propose a large dam on the Colorado.

DAVIS'S IDEA

H. T. Cory's victory over the Colorado was fragile and temporary. The river writhed in its bed, lashing out in periodic floods. Silt still piled up on the river bottom, raising the river's elevation and increasing the danger of greater floods. Silt still clogged the farmers' canals. The Salton Sea, 380 miles square, mocked the idea that small plans could govern the river.

Yet for a decade and more, no one seemed to learn the lesson. A jumble of tribal interest groups fought over the river: towns; counties; irrigation associations; farmer-entrepreneurs; absentee landowners; progressive politicians seeking small homesteads for World War veterans; Mexican bureaucrats and revolutionaries; Californians craving water for their cities; Arizonans suspicious of the Californians. All wanted their own scheme, and no scheme would serve all.

Soon after World War I, an engineer who knew the Colorado well stepped forward with a plan. He was Arthur Powell Davis, director of the U.S. Reclamation Service, established under President Roosevelt in 1902 as a branch of the Interior Department. No one since Davis's own uncle, John Wesley Powell, had proposed such a massive idea for the West.

Tall and athletic, with a startling crop of prematurely white hair, Davis first glimpsed the Colorado from a cliff in the Grand Canyon, on a trip with his uncle. From then on the river obsessed him. He rose through the ranks of western civil engineers, first as an expert in river hydraulics with the U.S. Geological Survey and the Reclamation Service; then as chief hydrographer on the Panama Canal; then back at Reclamation, where he became director in 1914, overseeing the construction of dams throughout the West.

Davis gained a reputation for unmatched knowledge and unshakable principle. A devoted conservationist of the Roosevelt school, he believed land and water were the common man's birthright, and they should be managed accordingly by the people's stewards in the federal government. "I have . . . little interest in conserving anything for the temporary benefit of a few 'dollar chasers,'" Davis said. "The broader principle of conserving all natural resources for the benefit of all persons is a cause for which, in common with many others, I should be glad to sacrifice anything." In short, when the nineteenth century's Jeffersonian faith in the family farm met the twentieth century's knowledge of what could be done with massive amounts of capital, concrete, and steel, the result was Arthur Powell Davis.

In its natural state, Davis said, the Colorado River was not much good for human use. (The argument that wilderness might be a good in itself would be little heard for many decades.) It fluctuated too much, running out of control in one season and dwindling to a trickle in another. Floods would keep doing damage. Silt would keep clogging irrigation canals and make the water unsuitable for drinking. Farmers and city-dwellers alike needed a steady, dependable, silt-free supply.

In some deep canyon of the Colorado, Davis said, one could build a very tall dam with a giant reservoir behind it. When floodwaters came racing down the river, the dam would stop them and the reservoir would store them. That would mean no more floods in the lower valley, and plenty of water for irrigation when the dry season came. Silt would sink to the bottom of the reservoir. New irrigation canals could be built without fear of siltation, and downstream drinking supplies would be clear.

Davis envisioned the Reclamation Service as the mastermind. But the project would require extraordinary cooperation between the federal government and the states. Even the Panama Canal wasn't a proper precedent for such a cooperative undertaking. Still more vexing was the question of who would pay for it. Davis's answer was a political master stroke. The dam, he said, could be built around a huge electrical generating plant, like the one at Niagara Falls. The plant would supply power to the rising cities of the West. The electricity not only would light those cities but supply them with water from the reservoir via canals and mountain-spanning aqueducts. The cities would pay for the power, and thus for the dam.

Davis's grand idea set in motion a dozen whirlwinds of political and engineering activity. In seven western state capitals, endless meetings, whispered conversations, and shouted threats ensued as

THE GREAT ENGINEER

The Colorado and Mississippi Rivers, oblivious to the human strivings along their courses, nevertheless helped to elect a president—Herbert Hoover, the only engineer to inhabit the White House and a tireless enthusiast for river development.

In the early decades of the twentieth century, Hoover gained fame as an engineer of genius, developing mines around the world; then as the "great humanitarian" who organized the feeding of the United States's European allies during the desperate last years and aftermath of World War I. This achievement propelled him into national politics, and he was eager for high office. "Just making money isn't enough," he confided to a friend. He longed to "get in the big game somewhere."

As secretary of commerce (1921–1928) under Presidents Warren G. Harding and Calvin Coolidge, Hoover conceived and promoted a sprawling vision of enhanced inland waterways that would cut freight costs and boost the national economy. For the Northeast and Midwest, this would mean a St. Lawrence River seaway; for the nation's midsection, flood control on the Mississippi and its tributaries; for the West, giant power- and flood-control developments on the Columbia and Colorado. Hoover called the Colorado his "favorite horsepower."

In 1921, the seven states that shared the river sent representatives to a new Colorado River Commission charged with working out terms for sharing the water. Hoover volunteered to act as chair, but it was a difficult job. Cooped up for eighteen days in a backcountry lodge near Santa Fe, New Mexico, the conferees included "more fractious elements," Hoover told a friend, "than any . . . I saw in Paris," referring to the peace conference at the conclusion of World War I. But the historian Norris Hundley says Hoover's pointed questions and shrewd management deserve much of the credit for what came to be called the Colorado Compact. The secretary's proposed compromise for water sharing was critical. Hoover called it "the first occasion when more than two states have under the direct provisions of the Constitution accomplished . . . the solution of interstate difficulties outside the Courts."

But the compact was only one step toward control of the river and only the beginning of Hoover's labors on the project's behalf. The legislatures of the participating states still had to approve it, then Congress. Hoover shuttled from state capital to state capital, lobbying and urging. He fought private-power magnates who wanted to develop the river themselves, or at least to keep Washington out of the game. "All this [power] will belong to the people," Hoover vowed, "developed by them, owned by them and for their benefit." When Arizona refused to sign the compact, Hoover made his rounds of the states again, crafting a deal by which the project could move forward without Arizona's approval.

In the midst of Hoover's campaign, the Mississippi tore through its levees in the great flood of 1927. Coolidge named Hoover to head federal flood relief, and he left Washington immediately by special train.

With the nation watching, he sped from town to town—to ninety-one communities in all—demanding facts and cutting red tape. Summoning his experience in wartime relief work, he set up tent cities, commandeered supplies, organized a rescue flotilla of some six hundred boats, and ordered a trainload of food be sent down from Chicago. With Coolidge's authorization, he coordinated the work of army engineers, local militias, the Coast Guard, the Weather Bureau, and the Red Cross.

He stopped short of a major military mobilization, believing local volunteers could do the work better. At every train station, he recalled, he would point a finger at town leaders and say: "A couple of thousand refugees are coming. They've got to have accommodations. Huts. Water-mains. Sewers. Streets. Dining-halls. Meals. Doctors. Everything. And you haven't got months to do it in. You haven't got weeks. You've got hours. That's my train."

"I suppose I could have called in the whole of the Army," he once said, "but what was the use? All I had to do was to call in Main Street itself. No other Main Street in the world could have done what the American Main Street did in the Mississippi flood."

The next year, Hoover defeated Governor Al Smith of New York for the presidency. His highly publicized role in the Mississippi flood had helped to reignite his chances for higher office, and his election, in turn, helped to rescue the Colorado dam bill from death by filibuster.

Hoover's name was affixed to the Colorado dam not once but twice. In 1931, with the honoree still in the White House, Congress approved the name "Hoover Dam." But after Hoover's defeat at the hands of Franklin Roosevelt the following year, Harold Ickes, the new secretary of the interior and a Hoover-hater of the first order, reversed that decision in favor of the original "Boulder Dam." In 1947, an act of Congress restored Hoover's name to the towering structure in Black Canyon.

Key visitors make up an inspection party for Diversion Tunnel No. 2 in 1932. From left to right: William Stringfellow, reservation police; Sims Ely, Boulder City manager; Henry Kaiser, from the board of directors of Six Companies, Inc.; Ray Wilbur, secretary of the interior; Warren Bechtel, president of Six Companies; W. L. Honnald, chairman of the Metropolitan Water District's engineering committee; S. D. Bechtel, from Six Companies board; W. P. Whitsett, president of the Metropolitan Water District; Frank Crowe, construction superintendent; and Glenn Boddell, chief reservation police officer.

politicians and bureaucrats dickered over how to share the promised water. Talks culminated at last in the Colorado Compact of 1922, which divided the water equally between an "upper basin" (Wyoming, Colorado, Utah, and New Mexico) and a "lower basin" (Arizona, California, Nevada). At the lonely Nevada-Arizona border, Reclamation surveyors scaled cliffs, drilled beneath the river, and recorded enough data to fill eight published volumes. They focused on two stretches of river, Boulder Canyon and Black Canyon. Both offered obstacles, but also the right combination of geologic and geographic characteristics.

From the time Davis first explained his idea to Congress, ten years passed before the essential pieces fell into place. In 1928, Congress passed and President Calvin Coolidge signed the prematurely named Boulder Canyon Project Act. Herbert Hoover, elected president that year, signed the enabling legislation in 1929.

The Contractors

Henry Kaiser was laying two hundred miles of highway across western Cuba when someone told him that Washington was inviting bids to build the biggest structure in the world. Chubby, bald, and jowly, Kaiser was forty-eight in 1930 and looked ten years older, but he did everything with a young man's urgency. In little more than a decade, without education or family wealth or even a background in building, he had become one of the most respected construction contractors in the West. He figured things out on the fly. He hired "people who are smarter than I am." The Kaiser Paving Company, founded at a time when men still built roads with shovels and mules, beat its deadlines by trying new technologies and efficiencies. Once, in the early 1920s, an engineer offered to show Kaiser an experimental construction machine of his own design. He called it an "earth-mover." For a little while Kaiser watched the tractor's buckets scrape up soil at the rate of ten cubic yards per minute. He offered to buy the patent on the spot. Then he offered the engineer a job. Every other contractor had seen the earthmover as a tool for doing a small job fast, the engineer recalled. Kaiser saw it as a tool for making a big job small.

The Cuba highway was itself a big job, Kaiser's biggest yet, in fact, and his first work overseas, probably the longest road in the Caribbean. That was good. But the biggest structure in the world—to a man of Kaiser's ambition that was so much better as to be irresistible. "I lay awake nights in a sweltering tent," Kaiser said later, "thinking about it over and over."

Soon, back in the United States, Kaiser strode into the San Francisco office of his friend and mentor, Warren Bechtel, the aging dean of West Coast construction. Kaiser didn't have the money to submit his own bid for the dam. But he told Bechtel their two companies might pull it off together.

Kaiser must have known his friend would hesitate. Bechtel still operated by the rough codes he had learned as a pick-and-shovel laborer in the early railroad camps of Oklahoma. The enormity of this project, with all its federal regulations and legal complexities, was a lot to swallow for a builder who believed that if you couldn't trust a man's spoken word there was no point in getting his signature. Together, he and Kaiser had built highways and a string of dams. But this dam was different.

"Henry," the older man said, "it sounds a little ambitious."

In fact it was the largest construction contract in U.S. history. The Reclamation Bureau (renamed in 1923) had published specifications that filled 100 pages and included 119 separate bid items, including 4.4 million cubic yards of concrete (more than all of the bureau's previous dams and irrigation projects combined); and 45 million pounds of pipe and steel. The contractors would have to dynamite and excavate four million cubic yards of rock. In a remote desert canyon with virtually vertical walls, they would have to build the world's biggest concrete factory; the nation's biggest power plant; a reservoir 115 miles long; and the world's tallest dam. Before they could do any of that, they would have to reroute the least predictable major river on the continent. And build a city for the workers in a desert. And put up a surety bond of $5 million. Scores of builders had read these specs already and said "No thanks."

But Kaiser knew Bechtel was not immune to the lure of a big idea, and he had persuasive talents that one day would get the better of Cabinet secretaries and presidents. "Henry is like a happy elephant," an associate said. "He smiles and leans against you. After a while you know there's nothing left to do but move in the direction he's pushing." Kaiser kept smiling, kept leaning, and Bechtel finally agreed to try.

Officials of Six Companies (right to left): H. W. Morrison, the head of Morrison-Knudson; Charles Shea, director of construction; Frank T. Crowe, chief engineer; and Henry Kaiser. An associate described Kaiser as "a happy elephant. He smiles and leans against you. After a while you know there's nothing left to do but move in the direction he's pushing."

The two men soon learned that even in combination they couldn't raise the necessary capital to back their bid. So they sought out allies, some of whom were already contemplating their own bids. They talked to desert-weathered builders like themselves: the Mormon brothers Edmund and William Wattis of Salt Lake City, born before the Civil War, schooled as teamsters on the Great Northern Railroad, builders of California's Hetch Hetchy Dam; the Wattises' ally, Harry Morrison, who started his career as an ax man for the Reclamation Bureau; Charlie Shea, a profane plumber who installed a water system under the hills around San Francisco Bay; Alan MacDonald, builder of sewers and skyscrapers up and down the West Coast, the only college-trained engineer in the group but a boss who persuaded recalcitrant workers with his fists. One by one these men made common cause, forming a construction monster they called Six Companies, Inc.

In February 1931, they began to calculate their bid. How to estimate a job projected to take seven years and more materials and equipment than any job in history? They studied the specs. They compared guesses. After long days of meetings in San Francisco, Henry Kaiser would drive all night to the dam site, sleep in his car for an hour, spend the next day inspecting the canyon, then roar back to San Francisco.

As the contractors' estimates began to converge, heads swung repeatedly to one man. Among the businessmen Kaiser may have been the driving force. But he would not actually build the dam. That job, all agreed, would fall to Frank T. Crowe.

Years before, on his honeymoon in New York City, Frank Crowe had left his bride alone for several hours while he watched a new kind of dump truck deposit coal into a small chute without a spill. As an engineering student at the University of Maine, he was so inspired by a Reclamation man's lectures about dam-building in the West that he signed up on the spot. He fell in love with the West and spent most of his life on the sites of rising dams, first as a Reclamation engineer, then as the indispensable man in Harry Morrison's construction firm.

Frank Crowe, chief engineer for the Hoover Dam, stood six foot six inches.

Crowe—tall and homely, with ears that stuck out—was equally effective as a solver of technical problems and as a leader of men. Workers trailed him from job to job. He loved construction sites, telling a reporter over the shriek of machinery: "I can't see why anybody would work in an office when he can be in a place like this." He didn't talk a lot. Few of his letters were longer than three sentences. When asked for an autobiographical sketch of one thousand words, he submitted forty. When Crowe was young, he had gone through the Colorado canyons with Arthur Powell Davis, who told him a great dam would be built there someday. "I was wild to build this dam," he said later. "I had spent my life in the river bottoms, and it meant a wonderful climax—the biggest dam ever built by anyone anywhere."

Crowe sifted the specifications again and again, teetering between the horns of the construction man's classic dilemma. If Six Companies bid too high, they would lose the job, probably to an eastern heavyweight such as the Arundel Corporation, which was better known to bankers and more experienced. With the construction industry withering amid the Great Depression, that was a grim prospect for the westerners. But if they got the job by bidding too low, they would lose their shirts and quite possibly their companies. Said one of the engineers: "We were all scared stiff."

Just two days before the scheduled opening of bids, the Six Companies men met to make their decision in a Salt Lake City hospital room where William Wattis was recovering from an illness. Crowe wheeled in a model of the dam, then walked the men through every item. They made tiny adjustments in his estimates, added a profit margin of 25 percent, and shook hands.

The Reclamation Bureau had sent specs to more than a hundred companies. In the bureau's jammed office in Denver on March 4, 1931, only five firms actually submitted sealed bids. The first two lacked the required bonds. The third envelope held Arundel's bid. It was for $53.9 million. The fourth came from Woods Brothers, a Nebraska firm—$58.6 million.

Raymond Walter, the Reclamation Bureau's chief engineer, opened the fifth envelope. He read aloud: "Six Companies, Inc., San Francisco, California, $48,890,955." Crowe's estimate was a slim $24,000 above the bureau's own.

The westerners of Six Companies got the job. What they would make of it—a brilliant victory over the Colorado or a bankrupting boondoggle—depended on the skill and speed of the men they hired. Crowe, now superintendent of construction, began to write a list of names he knew from building sites across the West—men who soon would receive a telegram saying simply: "OK for a job at Hoover Dam. F. T. Crowe."

The Boss

No one could erect a permanent dam smack in the current of the Colorado River. Before an ounce of concrete could be laid, the river had to be put somewhere else. This, of course, seemed a definition of the impossible.

Of the possible sites for the dam, Reclamation engineers had chosen Black Canyon. (Arthur Davis's first, tentative choice, Boulder Canyon, led to years of confusion about the dam's name.) Here the river cuts through a mass of purple-black rock that geologists call volcanic tuff. Over the ages, forces far down in the earth had pushed this mass upward. At the surface, the Colorado and its sharp grains of silt wore a groove in the rock. Year after year the rock rose and the river ran, one irresistible force pushing upward, the other downward. The eventual result was a crooked crevice a quarter mile deep. The planet took 12 million years to make this route for the Colorado. Frank Crowe had eighteen months to make a new one.

He had no choice but to send the river through the rock itself. Four great tunnels averaging four thousand feet in length had to be dug through the cliffs to divert the river around the site, leaving the riverbed empty and dry so that

A payment for one month's worth of construction.

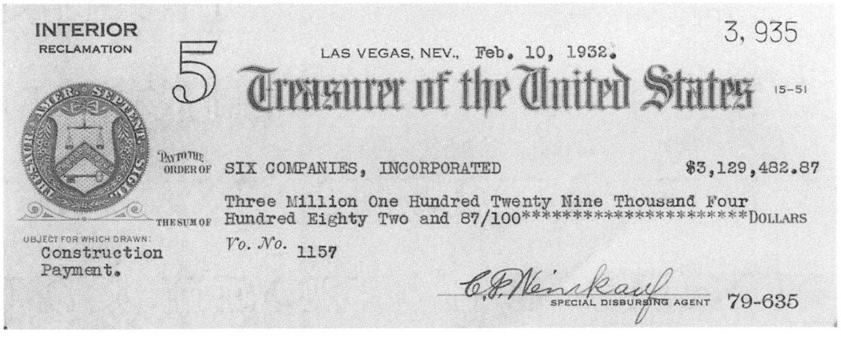
INTERIOR
RECLAMATION
3, 935
5
LAS VEGAS, NEV., Feb. 10, 1932.
Treasurer of the United States 15-51
PAY TO THE ORDER OF SIX COMPANIES, INCORPORATED $3,129,482.87
THE SUM OF Three Million One Hundred Twenty Nine Thousand Four Hundred Eighty Two and 87/100********************** DOLLARS
OBJECT FOR WHICH DRAWN: Construction Payment.
Vo. No. 1157
SPECIAL DISBURSING AGENT 79-635

Looking north through Black Canyon, June 21, 1932.

One of the four tunnels that would divert the river around the construction site.

construction crews could dig down to bedrock, then put up the dam and the power plant. A temporary blockage called a cofferdam would be built to steer the river into the tunnels.

In the spring of 1931, Crowe planned his schedule. To ensure quick progress, the Reclamation Bureau insisted that Six Companies finish the diversion tunnels by October 1, 1933, or pay a three-thousand-dollar fine for every day beyond the deadline. That gave Crowe two and a half years. But the river dictated a shorter schedule. It could only be diverted in late fall or early winter, when its flow was weakest; otherwise, flash floods could blast out the cofferdam and drown men working in the riverbed. So the tunnels actually had to be finished by the low-water period in the winter of 1932–33 or Six Companies would risk a massive financial penalty in only the first phase of the job. Crowe faced many tasks when he arrived at the site—laying roads into the canyon; arranging for power lines from California; building a town for workers; locating gravel sources; starting concrete plants; preparing the canyon walls. But the quick completion of the four great tunnels—each of them five stories high and as wide as a four-lane highway—loomed as the central problem.

For this job Crowe hired rock miners, most of them Irishmen who had blasted holes throughout the Mountain West to find copper, silver, zinc, and gold. He assigned them to his top assistant, a stocky, quick-thinking twenty-nine-year-old named Woody Williams. In each of the four cylindrical shafts, Williams had to choreograph an intricate ballet. Digging a tunnel through rock required three steps, endlessly repeated. First, miners had to drill a pattern of horizontal holes, some as long as twelve feet, into the tunnel's face. These holes would be filled with dynamite sticks and blasting caps, which then were detonated to shatter the surrounding rock. Other men would scoop out the shattered rock and truck it away, a process called mucking. Over and over, the crews would drill, blast, and muck, each round extending the tunnel a few yards further through the rock.

But it wasn't a matter of digging just one such hole. Each tunnel was so big it had to be dug in five separate sections. First, at the top, miners had to dig an "attic" tunnel, twelve feet square, to provide access for equipment and ventilation. Then, to either side of the "attic," they had to blast wedge-shaped openings called "wings" to complete the top of the circle. Then came the great hole called the "bench," fifty-plus feet wide and thirty feet high. Last, they had to dig out the bottom of the circle, called the "invert."

Drill . . . blast . . . muck. Each task required ninety men at a time in each tunnel and hundreds more in support. Men and machines had to move in and out of the cramped shafts at maximum speed. This was the key to meeting the tunnel deadline.

But the massive "bench" sections presented a terrible puzzle. They were simply too tall. For drillers to get at the upper face of the rock, a high scaffold was needed in the tunnel. Then, before blasting and mucking, the scaffold needed to be laboriously taken down and removed in a snarl of cables, apparatus, timbers, pipes, and people. It would be impossibly slow.

Studying the problem, Woody Williams perceived that a man holding a tool could not be the essential unit of tunnel construction. Rather, he must improvise a tool to hold men. Looking around at what he had on hand, Williams stripped a ten-ton army truck down to its chassis. Atop the chassis he welded a superstructure of steel pipes and wooden platforms that could hold thirty drills and men to work them. This automotive monstrosity, soon dubbed the "Williams Jumbo," could be backed up to one side or the other of the bench face for drilling, then moved to the other side for a repeat performance, then pulled out to allow for blasting and mucking.

Most of the men who built Hoover Dam were hired in this plain office in Las Vegas.

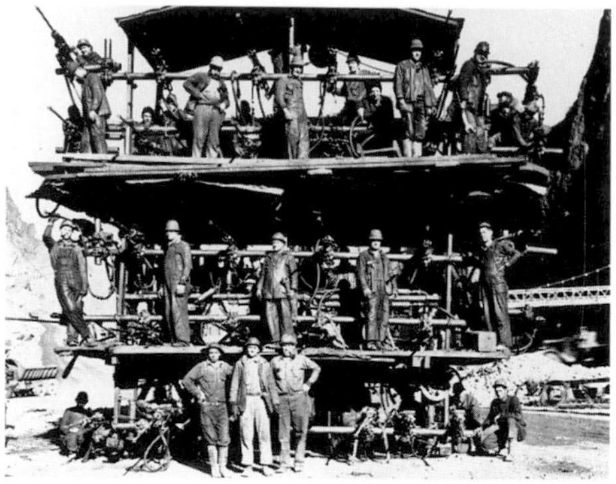

The Williams Jumbo, invented by Woody Williams for the Hoover Dam job, could hold thirty drills and the men to work them. The jumbo was used to drill the four diversion tunnels through the canyon walls of the Colorado River.

Williams's jumbo at work in its first operation.

Early work on one of the four diversion tunnels. Each was as wide as the Lincoln Tunnel in New York City.

The jumbo was tall but not so wide that it blocked other equipment in the shaft. With the jumbo, men could drill many holes at a time, and quickly.

The crews went at the rock in a mélange more intricate than any football play. Dodging the jumbos, goaded by shouting supervisors like the feared Red McCabe—"so tough he would fire a man for even looking like he was going to slow down," a worker said—the crews literally ran up and down the sloping shafts. Forty miners per shift rode the jumbos and drilled the holes. Forty chuck tenders fitted the drills with lengths of steel. Eight "nippers" fetched steel, water, or whatever else the miners needed. At the tunnel face, bathed in hot floodlights, the noise was too great for any communication but hand signals. When a set of holes was ready and tamped full of explosives, the jumbo was backed out, the charges blew smoke into the canyon, and the muckers rushed in with a bulldozer and a customized one-hundred-ton electric shovel to scoop the slag into fleets of waiting dump trucks. (The drivers of these trucks became notorious for their mad but time-saving practice of *backing* their empty trucks at high speeds up and down the steep, narrow roads of Black Canyon.)

The tunnel men worked against the deadline, against each other, and against other crews, competing for pride—more than 1,000 men toiling in round-the-clock shifts on 4 tunnels. They went fast. The record for one crew on one 8-hour shift was 46 feet of tunnel. On January 29, 1932, the muckers who reentered Tunnel Number 3 after a blast felt a draft against their faces from the far end of the tunnel. It meant they were the first to "hole out," completing their shaft, the shortest of the four at 3,560 feet, from one end to the other. Crews on the other tunnels soon caught up.

But a hole was not a tunnel. The crews still had to remove the "invert" from the holes' floors and trim jagged outcroppings from the walls, a job of several more weeks. Then the men had to line the holes with three-foot-thick sleeves of concrete strong enough to resist cavitation—the destructive suction of future floodwaters that could tear chunks out of the tunnel walls unless they were made smooth.

Very late one night, as an exhausted crew stood around debating how to pour concrete for one tricky section, a wide-awake voice pierced the murky air: "Who is holding up this pour?!" The men turned to look. It was Frank Crowe.

Hundreds of men now swarmed over the site at all hours, but it

The heat in the canyon often reached 120 degrees; in the tunnels it was sometimes as high as 140 degrees. Water boys, shown here at the Arizona spillway, carried sacks of water to the workers.

Plenty to eat for the dam worker in the boardinghouse cafeteria in Boulder City, the town created and built by Six Companies to house its five thousand workers and their families. Most of the company's accommodations were spartan.

remained very much Crowe's job. The workers regarded him as exceptionally demanding but fair, respectful of any hard worker and willing to get just as dirty as the grubbiest mucker. He walked the site with no hint of social superiority. One day a newly hired man was boarding the bus to the workers' quarters. He pointed out a bedraggled older man planted on a rock. "That poor old man is too tired to even get on the bus," he remarked. As the new man watched, a battered Buick pulled up. A second rumpled figure was at the wheel. The old man climbed in and the Buick rumbled away. The "poor old man," a veteran explained, was Charlie Shea, the millionaire contractor who, of all the Six Companies directors, knew the site best. The new man nodded, impressed. "And that was his chauffeur?" No, he was told. That was Frank Crowe.

As the tunnels neared completion, Crowe ordered the building

A worker goes "over the side" to drill on the rim of Black Canyon.

of a heavy timber bridge across the Colorado just upsteam from the yawning tunnel entrances. On November 13, 1932, 321 days ahead of schedule, dump trucks moved out on the bridge. Each dumped a load of rock in the river, then returned to the shore for another load. They were building the temporary dam to divert the Colorado into the tunnels. For most of the day and all that night, one load of rock splashed into the river every fifteen seconds. The current boiled against the rising wall but did not budge it. By 7:30 in the morning, the rising river was inching toward the portal of Tunnel Number 4. As Crowe and others watched, the river turned away from its old bed and entered the black tunnel.

In the canyon, still in deep morning shade, a man yelled: "She's taking it, boys! By God, she's taking it!"

The Wall

Not many places in North America are as brutal to the outdoor laborer as the canyonlands of the lower Colorado. Summer heat there is frightful, often exceeding 100 degrees Fahrenheit day after day, sometimes 120 degrees. In July 1931 the average low at night was 95, making sleep a torture. In daytime the black rock of the canyon became the walls of an inferno. In the diversion tunnels the temperature sometimes reached 140 degrees. Outside there were no trees and no shade. Many workers had to be in places too dangerous for water carriers to go; the result was an epidemic of dehydration. A doctor said men arrived at the nearest hospital "looking like they had been parboiled."

In the first year, living conditions for the men, some of whom had brought their families, were intolerable—except that they did tolerate them. President Hoover had ordered work to begin six months early as a relief measure for depression victims. That meant families had to live at the site before permanent quarters could be built, so they sweltered for months in abysmal, makeshift camps.

Frank Crowe lost twenty-seven pounds the first summer on the site. Yet Six Companies was a severe master. Under enormous financial and political pressure to work fast, the contractors seldom hesitated to put workers at risk if it meant saving time. Indeed, the gravest threat was manmade—the gasoline fumes in the diversion tunnels, which caused many men to stagger and sicken and some to die. When complaints about the gas fumes went to court, Six Companies fought them fiercely rather than switch to electric engines.

Under pressure, the company made conditions better, especially when workers' quarters at the new Boulder City were completed. Even then, the rock-strewn cliffs and powerful machinery made Black Canyon a very dangerous place. Six Companies' safety signs reminded workers that DEATH IS SO PERMANENT, and for ninety-six workers (not counting the uncertain number of men who died from inhaling gas fumes) the warning came true. Fatalities at Hoover Dam ran much higher than for comparable projects such as the Brooklyn Bridge, where twenty died; the Golden Gate Bridge, where eleven died; and the Empire State Building, where only five died. "It is no wonder that they are forced into a kind of bravado that goes with the game," a journalist wrote in the *New York Times,* "that they take to minimizing the risks and the inhuman enormity of the job . . . that they adopt a tone of sardonic superiority to dangers, painting in huge white letters on the door of a [blasting] powder house, 'God Bless Our Home.'"

Five thousand workers stayed on because the depression offered them little choice. Some were experienced miners and construction men. Many more were farmers, clerks, teachers, and professionals who had run out of options. One was an unemployed trumpet player. They worked hard and fast out of pride—the slogan of men who followed Crowe from job to job was "If it's too tough for anyone else, it's just right for us"—or out of the simple knowledge that they would be fired if they didn't.

The diversion of the river through the canyon walls ranks as one of history's great feats of engineering. Yet one measure of the immensity of Hoover Dam is that the tunnels were merely a preliminary to the main task at hand.

For years, Reclamation engineers had been pondering what sort of wall would not only halt the great river in its ancient bed and hold back its immense pressure—some 45,000 pounds per square foot at the dam's base. They must also block this force of nature beyond any possibility of failure. Anything less would court catastrophe. After considering various designs, they chose the form known as an arch-gravity dam. It would take the shape of an inverted wedge; that is, a wall thick at the bottom and thin at the top, curving into the cliffs on either side. With this ingenious shape, the wall would play a kind of trick on the river, transferring the water's enormous weight through the concrete to the canyon. In other words, the water would jam the dam into place, compressing the concrete, while the dam's

STIFFS

"The [Hoover] Dam worker of 1933 is a national type of some importance. He is a tough itinerant American—the 'construction stiff.' His average age is thirty-three. His average wage is sixty-eight cents an hour. He is taller and heavier than the average U.S. soldier, runs a greater risk of losing his life, and has passed a more drastic physical examination. He has been in most of the states of the Union and can find his way in a dozen different kinds of unskilled and semi-skilled labor—a hoist in a Pennsylvania coal mine, a saw in Oregon, a shovel on a dozen road jobs. He has boiled a string of mules in Bluejacket, Oklahoma—followed a pipe line as it crept across a prairie, a few yards a day, toward a town invisible behind a hill range. He is inured to ceaseless, frightful heat—and fearful cold, too, for that matter. Four or five of him in an old car can always get to a row of lights on Saturday night and if some fourflusher cops his roll or his girl it may be a fight or a laugh—what's the difference? He has earned $10 a day roughnecking on top of 110-foot oil rigs, driven a steam shovel, been slashed in a dance-hall fight, thought a lot about getting married. He is sentimental, moody, and literate; he does not believe he will ever be anything better than what he is, and isn't trying, regardless of the schoolbooks, the adage to 'make your spare time pay,' and the example of Abe Lincoln. He leaves some money every week or so in block sixteen, Las Vegas (legalized prostitution), but has enough left to send a money order to somebody somewhere once a month. He shares the universal superstition of miners that if a woman ever walks into a tunnel where you are working you'd better get out quick because there's going to be a cave-in. He keeps washed. He smokes a pack of cigarettes a shift. When he travels, he rides freights. He knows how to live in jungles, but has never begged. The most he ever had in his life was $5,000 after the pipe-line job but he hung it on a wrong deal and lost it. He likes hunting better than baseball, horse racing better than either. He'll pick a grudge, or smell bad luck, mosey out and hit the road or the rails, but while he works he is inspired with a devil of loyalty, shrewdness, and skill. He wears Friendly Five shoes, and sleeps seven hours a day. He is the man, as much as General Superintendent [Frank] Crowe and U.S. Engineer-in-Charge [Walker] Young, who is putting up this dam faster than anyone thought it could possibly be done."

—*Fortune,* September 1933

Hoover Dam near completion.

"I feel like hell," Crowe said. "I'm looking for a job. . . . I've got to find a dam to build somewhere." He went straight to work on Parker Dam, downriver from Hoover, then two smaller dams in Colorado, and finally the soaring Shasta Dam on the Sacramento River, centerpiece of California's massive Central Valley Project. In 1945, when Crowe was sixty-three, the War Department asked him to direct postwar reconstruction in Germany, but his doctor told him his body wasn't up to it. He bought a ranch near Shasta Dam and died there in 1946. Many years later, when astronauts circled the earth and sent back pictures, someone pointed out that Frank Crowe was one of the few figures in history whose handiwork could be seen from outer space. Of course, the same could be said about Charles Rockwood and the Salton Sea.

The Symbol

It is easy enough to enumerate the tangible benefits of Hoover Dam. It became the keystone in a system of dams and canals that prevents floods; irrigates luxuriant, year-round farmland across southern California and Arizona, including the magnificent Imperial Valley; gives clear water to metropolitan Los Angeles and San Diego; and generates electrical power for Las Vegas and most of southern California. The population and industry of the modern Southwest are its offspring. The Allied victory in World War II owed much to ships and aircraft built in factories powered by Hoover Dam.

Yet just as the dam had its roots in the American myth of the garden, it loomed as large in the realm of symbolism as in the practical world of power lines and irrigation valves. Its career as a symbol began even before it was finished, when Americans looked to it for reassurance that their nation, staggered by the Great Depression, could still achieve great things. As its success became obvious, less developed countries pointed to the dam as a sign of what could be done to harness their own natural resources. In the United States, the dam stood as a powerful argument in favor of John Wesley Powell's vision of a mighty democratic technology. The details were not as Powell had imagined. But his central idea—that the community, not private interests, should shepherd land and water for the greatest good of the greatest number—prevailed.

Born in the first wave of American conservation, the dam took on an evil tarnish during the second wave, which began in the 1960s and continues today. As the journalist John McPhee put it, many environmentalists and wilderness advocates see dams—and Hoover, though no longer the biggest, remains *the* dam of dams—as "metaphysically sinister . . . because rivers are the ultimate metaphors of existence, and dams destroy rivers." There's no denying that the Colorado is now "a push-button river," computerized and controlled. And Frank Crowe was right; the dam will stand for centuries. But for the Colorado that's not such a long time.

Wilderness advocates may despise the dam, but at least they see it in monumental terms. A stranger view is that of the casual tourist for whom the dam is only a brief diversion from the frantic games of Las Vegas. Those who watched the dam rise could never see it that way. For them it implied more than material benefit. It seemed to create some new category of human endeavor. Their response to the dam was pride but also awe. The British writer J. B. Priestley looked into the canyon and saw "a new world . . . of titanic communal enterprises." Some similar emotion struck the journalist Theo White, who spent several weeks at the site to write a story for *Fortune* magazine. He described his last visit to the dam: "High up on the rim, in the late afternoon, I sit to watch it. It is a beautiful, tantalizing thing. It is complex. It has a meaning, not to be grasped in weeks, or perhaps years. . . . The sun sets. Long shadows climb down into the canyon. Crimson rock turns maroon. . . . Workers are scarcely visible. The thing loses scale. A steady hum, quiet and muffled, comes up between the walls, ripped violently by an occasional whistle from the high lines. I stay on, fascinated. I stare at the thing trying to comprehend it, to fix it forever in mind's eye. I have been inspired and provoked in the weeks I have tried to know it. And now, the minds that invented it, the bodies that are building it, the complexity and spirit of it, the love of it which men feel—all, all bewilder me."

Perhaps the elusive meaning White groped for was that Hoover Dam, more than any other single artifact of technology, seemed to destabilize the limits of what humans could do. That was cause for solemnity as much as for pride.

A pipe section is lowered into place at the entrance of the penstock tunnel, which is used to bring water out of the reservoir and take it to the turbines of the hydroelectric power plant.

2 POWER

"The next transformation would occur in a realm of things too small to be seen or heard, where electrons move along copper wires."

People of the nineteenth century associated great works of engineering with size and clamor. After all, their world was transformed by canals and bridges, by oceangoing steamships and monster locomotives. It took the revolutionary mind of Thomas Edison to perceive that the next transformation would occur in a realm of things too small to be seen or heard, where electrons moved along copper wires.

We remember Edison chiefly as an inventor of clever gadgets—the electric lightbulb, the phonograph, the motion picture. However ingenious and important those inventions, our fading picture of him as a gadgeteer falls far short of capturing the breadth of his achievements. In fact, Edison envisioned and brought to birth the entire system of electricity that powers the modern world. His lamp was only the most visible component of a much larger invention—the lighting district he constructed in lower Manhattan, which was the precursor of an electrified world.

Writers of the day portrayed Edison as the quintessential lone inventor. But his contribution as an engineer was more imoprtant than that. He created and led the first team devoted to technological research and development. With science as their guide and the mass market as their goal, the workers in Edison's Menlo Park laboratory—the first electrical engineers—not only created an industry from whole cloth, but developed an approach to technology that would shape modern society. Like the work of Edison's contemporaries, Henry Ford and Frederick Winslow Taylor, the business conducted at Menlo Park exemplified the historian Thomas Hughes's definition of modern technology: "the effort to organize the world for problem solving so that goods and services can be invented, developed, produced, and used." That's also a fair definition of the society that the United States and the other industrial nations carried with them into the twenty-first century.

Edison's vision of an electrified America can be measured by the many years that were required to make it real. That work was led by his most important disciple, Samuel Insull, a genius of a different breed. After a decade as Edison's chief lieutenant,

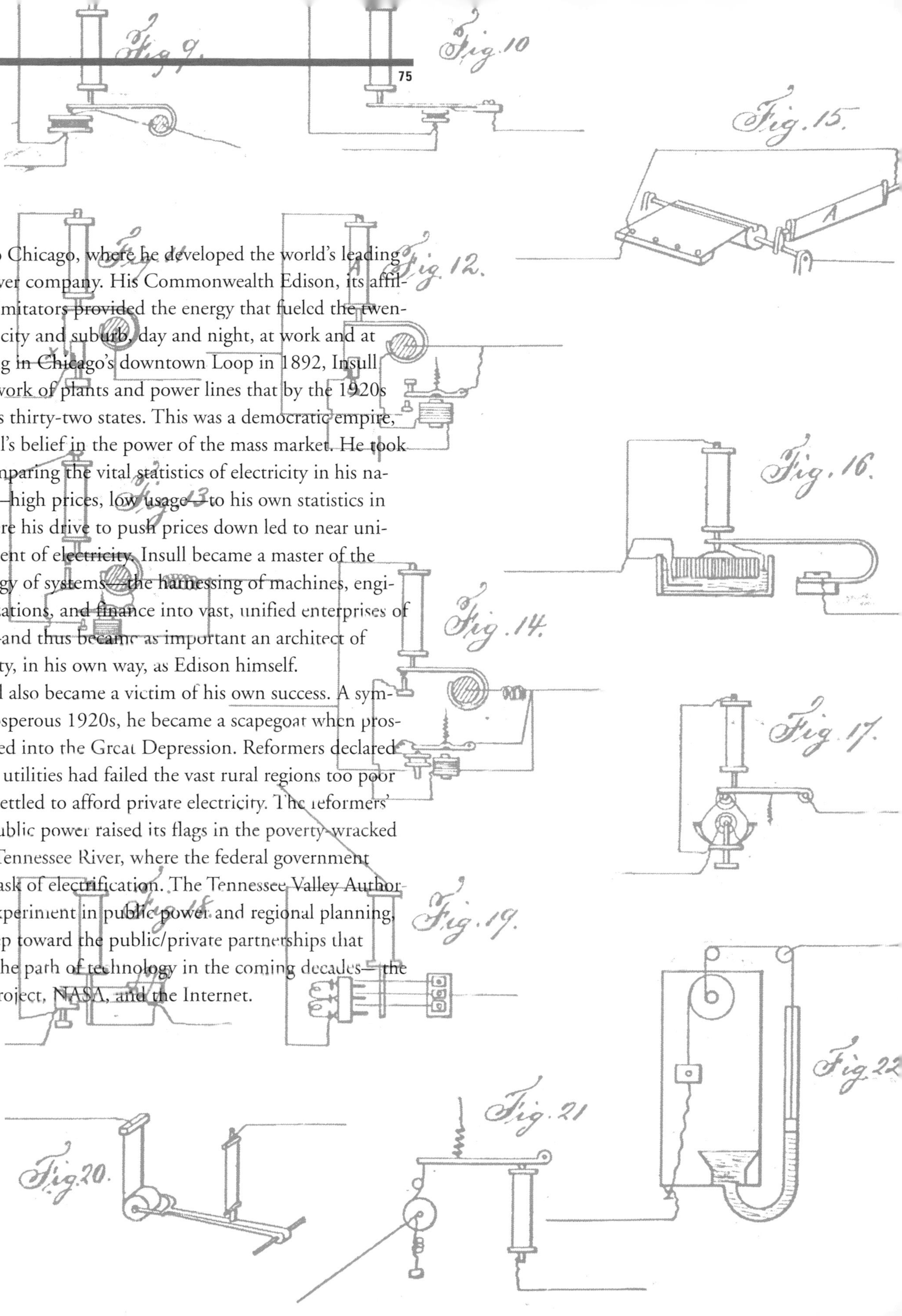

Insull went to Chicago, where he developed the world's leading light-and-power company. His Commonwealth Edison, its affiliates, and its imitators provided the energy that fueled the twentieth-century city and suburb, day and night, at work and at home. Starting in Chicago's downtown Loop in 1892, Insull created a network of plants and power lines that by the 1920s reached across thirty-two states. This was a democratic empire, built on Insull's belief in the power of the mass market. He took delight in comparing the vital statistics of electricity in his native London—high prices, low usage—to his own statistics in Chicago, where his drive to push prices down led to near universal enjoyment of electricity. Insull became a master of the new technology of systems—the harnessing of machines, engineers, organizations, and finance into vast, unified enterprises of production—and thus became as important an architect of modern society, in his own way, as Edison himself.

But Insull also became a victim of his own success. A symbol of the prosperous 1920s, he became a scapegoat when prosperity collapsed into the Great Depression. Reformers declared that the great utilities had failed the vast rural regions too poor and sparsely settled to afford private electricity. The reformers' crusade for public power raised its flags in the poverty-wracked valley of the Tennessee River, where the federal government took up the task of electrification. The Tennessee Valley Authority, a grand experiment in public power and regional planning, was a long step toward the public/private partnerships that would mark the path of technology in the coming decades—the Manhattan Project, NASA, and the Internet.

CHAPTER THREE

EDISON'S LIGHT

A charcoal sketch drawn nine months before the opening of Edison's Pearl Street Station.

Is it a fact—or have I dreamt it—that, by means of electricity, the world of matter has become a great nerve, vibrating thousands of miles in a breathless point of time?

— *Nathaniel Hawthorne*
The House of the Seven Gables

In 1879, a copper wire connected the two telephones in Menlo Park, New Jersey, a crossroads village twenty-five miles southwest of New York City. One sat in the home of the inventor Thomas Edison, on Christie Street. The other was in Edison's laboratory a hundred yards up the hill, a two-story frame building that looked like a good-sized country church, as wide and tall as a house but as long as a barn. At the dinner hour, one of Edison's aides would pick up the receiver in the lab and hear Mary Stilwell Edison ask when her husband would be home. The aide would check, then return with the same answer as the night before: "He'll be along presently, Mrs. Edison." Then Mrs. Edison would wait as her husband worked on, forgetting to go home until long after the dishes were put away and his children asleep.

For Edison was the head of two families, and only one drew his full attention. The first—his sad young wife and their two children, daughter Marion (nicknamed Dot) and Thomas Alva, Jr. (nicknamed Dash)—lived in the fine clapboard house he had bought in 1876. The members of his second family were sharp and industrious young men, each chosen for some skill or talent that augmented Edison's powers of creation. Every day and night some fifteen to fifty of them—the number varied according to the tasks at hand, with much turnover—worked in industrial noise and chemical stink, hammering, assembling, adjusting, testing, and experimenting with every implement and substance one could imagine.

One hour would find Edison checking on his projects, each of them pursued by a separate assistant but all the product of Edison's mind. The next hour might find him napping in a cubbyhole under

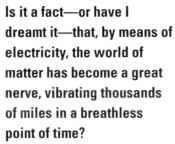

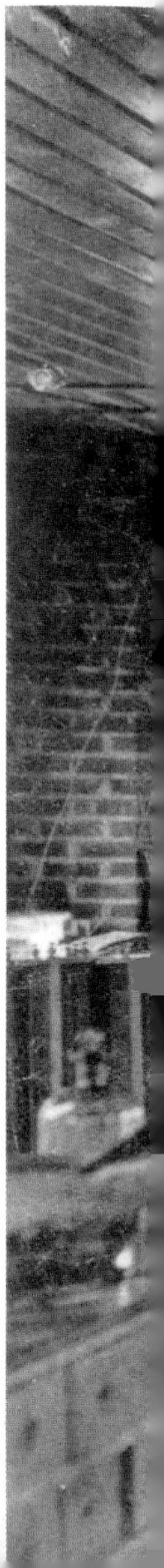

the stairs, or sprawled on the floor amid scattered books, pursuing some point of science, or talking John Kruesi, his machine-maker, or Charles Batchelor, who had served as Edison's chief assistant for five years. Although Edison grappled with the equipment as often as his aides did, one biographer says that his hands, though perpetually dirty, "were soft and beautifully formed—the hands of an artist, of a man who imagined things, not those of a workman or a craftsman."

The big room crackled with stories, gibes, and practical jokes, many of them perpetrated by the boss himself, who never outgrew his adolescent sense of humor. Though uncomfortable in social settings, Edison enjoyed the company of workingmen. He told stories well and liked to laugh hard. His skin was pasty pale from indoor labor. His hair was a mop, and he didn't bathe often. His dress bor-

Edison's employees posed for this 1880 group portrait at the Menlo Park laboratory, a second home to them. Edison is seated in the middle wearing a skull-cap. The lightbulbs hanging from the ceiling are screwed into old gas fixtures.

Mary Edison was Edison's first wife and the mother of his first two children, whose nicknames, Dot and Dash, were inspired by Edison's profit-making work in telegraphy. Mary suffered from chronic depression exacerbated by Edison's neglect, and she died in 1884, two years after his success at Pearl Street.

Lab assistant John Kruesi was a top experimental machinist at Menlo Park.

Charles Batchelor was one of Edison's closest friends and his most important laboratory assistant, the only one to attend Edison's bachelor party in 1886.

dered on the slovenly; his rich New York backers had to advance him money to buy a decent suit for business meetings.

Though he wasn't big or loud, he tended to dominate any room. In print, his claims looked like braggadocio, and he was undoubtedly an egoist of the first rank. Yet he won the respect of many intelligent and accomplished people, infecting them with his intensity and enthusiasm. "Edison is always absolutely himself," a visitor said, and "possessed by the joie de vivre." He expected his workmen to be as obsessed with his goals as he was, which meant, among other sacrifices, that they usually worked all night and seldom saw anyone but each other. Many had no taste for the regimen and quit. Those who stayed became Edison devotees. At midnight or later, they would gather at the hearth for pie or smoked herring on crackers. Someone would play the organ at the end of the room, and they all would sing—sometimes "We Won't Go Home Until Morning," other times "Over the Hill to the Poor House." Then they would go back to work.

Occasionally Edison submitted to a meeting with the sober men who bankrolled this nest of inventive endeavor. In private he called these men his "leaden collar," and he restricted dealings with them to a minimum. He had built his little kingdom precisely to free his mind from business distractions. His true business, he knew, was to invent, and the more time he spent in this yeasty environment of talk, thought, and experimentation, the more inventing he would do. Here he could follow his footloose curiosity, knowing it often led him to unexpected treasures of insight. With this method he had invented a stock ticker, improved telegraphs, a telephone receiver, and most recently the phonograph, his most original and remarkable invention to date. Reporters were calling him the "Wizard of Menlo Park" and following his projects closely.

"Look," he told one of them, "I start here"—poking a finger at a point in the air—"with the intention of going there"—drawing an invisible line—"but when I have arrived part way in my straight line, I meet with a phenomenon, and it leads me off in another direction—to something totally unexpected."

In the summer of 1878 Edison was racing along a line toward the perfection of the phonograph. Then events interceded that jolted him off that line into the most important new direction of his career.

The first occurred far from Menlo Park, on a western railroad vacation. Edison, exhausted after many months without a break, had

Edison in 1888, after several days without sleep.

THE WIZARD'S WORKSHOP

Henry Ford was a student, fan, and friend of Thomas Edison. His first job as an engineer was for the Edison Illuminating Company in Detroit in 1891, and he collected Edison materials for years. In 1927, when Ford oversaw plans for a museum in his name in Dearborn, Michigan, he brought together a vast amount of Americana and technological artifacts, with a large Edison collection at its core. In an early nod to historic preservation, buildings were purchased and relocated there, and replicas of others were constructed meticulously, including several from Edison's Menlo Park compound. The resulting village was dedicated in 1929 and opened to the public in 1933. Ford named it the Edison Institute, though it is now better known as the Henry Ford Museum and Greenfield Village.

Edison's machine shop.

Edison kept more than twenty-five hundred chemicals on hand. He had the most well-equipped private laboratory in America.

Edison's machine shop was crucial to his operation. Here his assistants produced the many items of iron and steel required by his laboratory for experimentation and production, including lathes, drills, and planers.

Edison's glassblowers fashioned glass bulbs for lighting, glass tubing, bottles, and jars for work at the Menlo Park lab.

A nineteenth-century copy machine. Edison's cyanotype machine made multiple copies of his drawings and plans by exposing chemically treated paper to the sun.

Edison's work area.

Francis Upton became a key assistant to Edison during the search for a practical incandescent lamp. Edison, aware of the quiet young physicist's roots in the New England aristocracy, so different from his own hard-scrabble background in the upper Midwest, quickly nicknamed Upton "Culture." But the inventor valued Upton for skills in methodical experimentation and math that compensated for his own shortcomings.

ing the flow of the current, he would not need so much current. Less current meant thinner conductors, and thus less money spent on both current *and* conductors. This insight, perhaps as much as his great gift for perseverance, kept the inventor optimistic in the winter of 1878–79.

He was sure his idea spelled ultimate success. But for the moment it brought him no closer to finding the right substance for his burner—a substance that would a) offer high resistance to a current; b) glow strongly but safely; and c) not burn or melt at high temperature.

With snow covering Menlo Park, Edison's jaunty claims of the previous September were echoing again in the press, this time with accompanying catcalls. "Edison is not a humbug," wrote the humor magazine *Puck* in a mock defense. "He is a type of man common enough in this country—a smart, persevering, sanguine, ignorant, show-off American. He can do a great deal and he thinks he can do everything." A British newspaper sneered: "All anxiety concerning the Edison light may be put to one side. It is certainly not going to take the place of gas."

The workers at Menlo Park were in no position to prove otherwise. They needed time. They couldn't simply keep trying one contraption after another in hopes of stumbling on one that worked. They had to dig deeper. If they lacked the tools to discover more about what electricity *was,* they at least could study what it *did* more closely. So they tested dozens of substances shaped into various forms—wires, foils, sticks, buttons—and carefully recorded all results. They coated burners with material that might retard the burners' self-destruction. Most important, Edison began to study experimental burners *after* they had disintegrated. The lens of a microscope revealed fissures in the ruined platinum and other metals. Pondering this, he theorized that the heat was causing tiny pockets of gas within the metal to expand. So Edison began a new set of experiments, this time using a sophisticated vacuum pump. He found that if he slowly heated a burner inside a glass globe to a temperature lower than the level of incandescence, he could force the gases out of their hidden pores. Meanwhile, the vacuum pump would suck the air and the gases out of the globe. This process left the burner much harder; it could then be heated to a much higher temperature without disintegrating.

The vacuum experiments were painstaking, involving hours of

delicate labor and repeated frustrations. "Just consider this," Edison told a visitor. "We have an almost infinitesimal filament heated to a degree which it is difficult to comprehend, and it is in a vacuum under conditions of which we are wholly ignorant. You cannot use your eyes to help you, and really know nothing of what is going on in that tiny bulb." One day Edison used an improvised arc light to heat the test materials. The light it gave off was so brilliant it seared his eyes. But he went at it from 3 o'clock in the afternoon until 10 o'clock that night, when the "pains of hell" forced him to quit. The pain in his head continued until 4 A.M., when a dose of morphine allowed him to sleep. But the vacuum experiments yielded promising results. In some tests of the hardened burners, the light leaped to twenty-five candlepower and lasted for many minutes.

This was good progress. But compared to the promises of six months earlier, it seemed barely a beginning, and the investors were asking Grosvenor Lowrey just what was going on at Menlo Park. When, at a board meeting, several of them took the absent inventor to task, Lowrey leaped to his defense, arguing that "no great end could be obtained without considerable doubt and tribulation." J. P. Morgan sat silently through Lowrey's speech. Then he signaled with a few words that he saw no need to pull Edison's plug. Yet. But doubts were mounting. When Dr. Henry Morton, head of the prestigious Stevens Institute and an authority on electricity, predicted Edison's scheme would come to nothing, Edison declared he would soon erect a statue of Morton in the yard outside the lab and shine an electric light on its countenance forevermore.

In the Menlo Park shops, work followed the preindustrial pattern of master and craftsman. Each man labored at his own pace, taking a nap or working late as he saw fit. Raucous jokes and insults were not only tolerated but encouraged. Electric shocks were administered routinely, to great hilarity. It was how Edison wanted it. The process of invention demanded both flexibility and perseverance, and he believed those traits flourished in a loose and jocular atmosphere.

But in the summer of 1879, the pressure of investors and Edison's own overwrought promises drove him into frenzies of work, and he drove his assistants in turn, often "to the limit of human endurance," one of them recalled. His intensity was both inspiration and burden. Francis Upton said his boss "could never understand the

The new Edison lamp as drawn by Francis Upton: "I shed the light of my shining countenance for $15,000 per share."

The first known photo of a working incandescent lamp. An Edison assistant looks on. Although its light was so radiant that it was painful for Edison to look at, he studied it for seven hours, until "the pains of hell" made him retire for the evening.

The dynamo nicknamed "Long-Legged Mary Ann" by Edison's lab assistants. The long magnets were a crucial improvement, increasing the efficiency of the dynamo to 90 percent. Until this design, conventional scientific wisdom held that the maximum theoretical efficiency of a dynamo was 50 percent.

limitations of the strength of other men because his own mental and physical endurance seemed to be without limit."

Under this pressure results began to emerge. The men working the vacuum pumps, which Edison had redesigned and improved, got more and more air out of the glass globes. A teenage German glassblower named Ludwig Boehm, recruited through a classified ad, proved very helpful in making better vacuum bulbs. By September the team could make a vacuum of one one-hundredth of an atmosphere. After a few more weeks it was one one-millionth.

Weeks of experiments had led to a better dynamo. It converted steam power to electric power far more efficiently than William Wallace's, even to the point that it could generate electricity cheaply enough to make business sense. The machine's most prominent feature was a pair of five-foot-long cylindrical magnets that rose straight up from the floor. These thick poles, with their accompanying apparatus on the floor, reminded someone in the shop of a recumbent female, her legs raised to welcome an admirer. So the dynamo was nicknamed the "Long-Legged Mary Ann"—"in honor," according to one Edison biographer, "of a girl who occasionally visited the lab." (Victorian newspaper editors converted the phrase to "Long-*Waisted* Mary Ann.")

Edison still was in desperate need of an effective, long-lasting burner. There was lots of carbon around the lab that year, since Edison, in addition to inventing, was overseeing the manufacture of carbon buttons to sell for his telephone receivers. According to the earliest account, Edison was idly rolling a glob of carbon lampblack between his fingers when it occurred to him that carbon might stand up well in a high vacuum.

On October 20, Edison charged a carbon stick in a vacuum bulb. The stick glowed at 40 candlepower, stronger than any platinum burner, and it didn't consume itself. Still, its resistance was low. Edison put several men to work making hair-thin reeds of carbon, several inches long, and twisting these reeds, or "filaments," into coiled spirals. But these broke too easily.

On October 21–22, Edison and Charles Batchelor tried a series of filaments including a one-inch piece of straight cotton thread. Late that night, they rubbed the thread with a mixture of carbon and tar, baked it, then carried it to the glassblower's shed behind the lab, where it could be inserted in an evacuated bulb. The thread broke. They tried again. Batchelor fitted the fragment into a new globe,

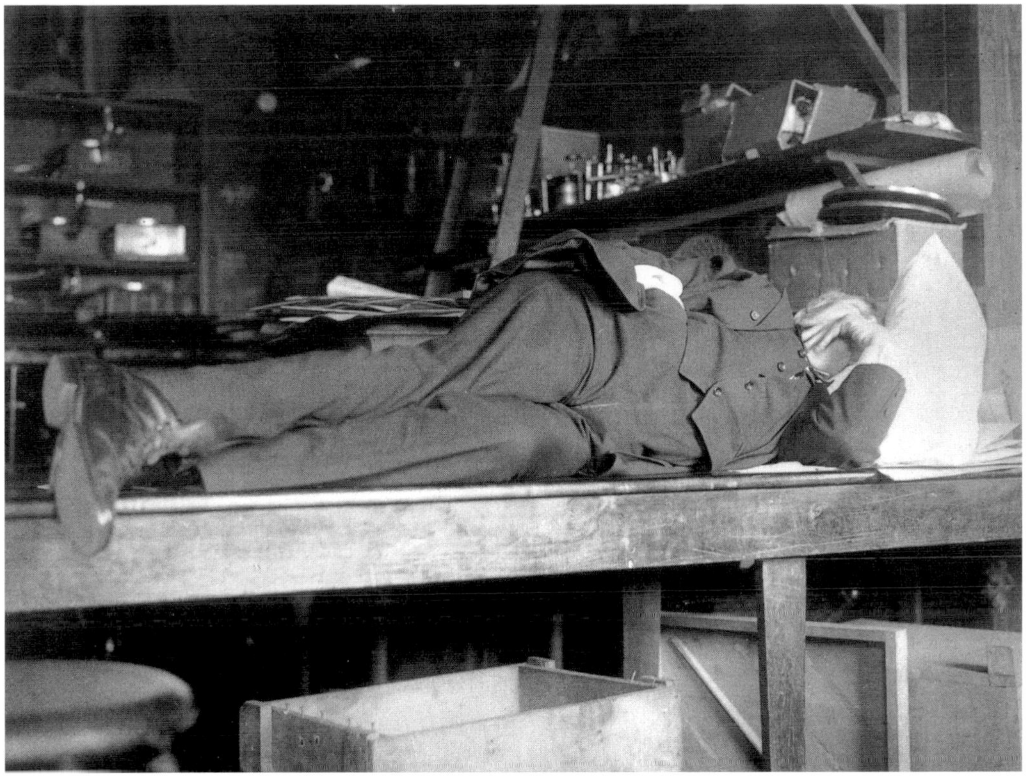

Edison's absorption in his work caused him to sleep many nights in the lab, even when he was an older man, as depicted in this 1924 photograph.

painstakingly evacuated the globe, and connected the filament to the copper wires. His face was illuminated by what he called "an elegant light equal to twenty-two candles." They kept at it, trying different kinds of thread and paper with various coatings. At 1:30 A.M. they tried a simple thread with no coating at all. It stayed lit until late the next afternoon, with a high level of resistance. Marshall Fox, the *New York Herald* reporter who had covered the story from the beginning, compared the glow to "the mellow sunset of an Italian autumn . . . a little globe of sunshine."

But the cotton was so fragile. Edison needed something with a similar structure, but stronger. Since the cotton filament was organic, he tried fragments of other plants—spruce, hickory, maple, cork, coconut hair, coconut shell, bagging, baywood, boxwood, cedar shavings, celluloid, fish line, flax, plumbago, punk, twine. He even electrified a whisker plucked from an assistant's beard. When he tried a filament made of tough cardboard, the lamp stayed lit for more than a week.

On November 4, 1879, with words that barely hid his exultation, he applied for a U.S. patent on a carbon-filament lamp: "I have

Ludwig Boehm, recruited by Edison through a classified ad, used his glass-blowing abilities to make better vacuum bulbs, a crucial element in the development of the electric lamp.

The Menlo Park lab, c. 1880. Edison saw his lab as a vehicle to serve his own creative vision, saying, "I am not in the habit of asking my assistants for ideas. I generally have all the ideas I want. The difficulty lies in judging what is the best idea to carry out."

discovered that even a cotton thread, properly carbonized and placed in sealed glass bulbs, exhausted to one-millionth of an atmosphere, offers from one hundred to five hundred ohms' resistance to the passage of the current and that it is absolutely stable at a very high temperature." But for once he wanted no publicity. He wanted the lamp to last hundreds more hours before he declared victory.

As the rejuvenated staff sprinted through more trials all through November, trying to improve on their fragile success, derision in the press continued. A scientist writing to *Scientific American* declared it to be "almost a public calamity if Mr. Edison should employ his great talent on such a puerility."

In spite of Edison's sudden conversion to secrecy, rumors seeped out. For a few weeks a strange war was waged between belief and disbelief. On December 21, breaking Edison's embargo on news, Marshall Fox published a story covering the entire front page of the *Herald:* "The Great Inventor's Triumph in Electric Illumination." But the papers had made such claims before, and the scientists remained dubious. The doleful Professor Morton, of the Stevens Institute, called the whole thing "a fraud upon the public."

As the Menlo Parkers strung wires through the Edison compound and the six nearby houses, people came out from the city at night to see for themselves. They went back with stories of a village glowing with a steady, yellow-white light that no one had ever seen before.

All doubts were put to rest on New Year's Eve 1879. That night, trains from New York, Newark, and Philadelphia disgorged some three thousand people at the Menlo Park station, and as 1880 dawned, people regarded each other in the first light of a new age.

"ALL I PROMISED"

"There is a wide difference between completing an invention and putting the manufactured article on the market," Edison asserted testily as the success at Menlo Park provoked renewed calls for quick results in New York. "The public, especially the public of journalism, stubbornly refuse to recognize this difference."

Of course, he had no one but himself to blame for heightened expectations, and the tasks he faced now, in invading Manhattan,

"A MATERIAL SO PRECIOUS . . ."

When Edison discovered that carbonized paper would glow for many hours in an evacuated lamp, he was delighted but hardly satisfied. "Somewhere in God's mighty workshop," he told a reporter, "there is a dense woody growth, with fibers almost geometrically parallel and no pith, from which we can make the filament the world needs." He needed it, too. He was facing patent-infringement lawsuits over the use of carbonized paper.

"If Edison had a needle to find in a haystack," said Nikola Tesla, the great electrical scientist who worked for Edison in 1882–83, "he would proceed at once with the diligence of the bee to examine straw after straw until he found the object of his search." To a pure scientist like Tesla, this dogged reliance on cut-and-try methodology was an abominable time-waster. "I was a sorry witness of such doings," he said, "knowing that a little theory and calculation would have saved him ninety percent of his labor."

But Edison craved the adventure of the search. In the spirit of the Jules Verne novels he loved, with their fantastic tales of jungle journeys and ocean expeditions, he conjured up his own scientific treasure hunts and dispatched surrogates around the globe. Even if they turned up nothing useful, he figured, they fed the newspapers' hunger for Edisonian copy at times when the inventor's own work was dragging. The papers called them "dauntless knights of civilization," searching for a modern Grail, "a material so precious that jealous nature had hidden it in her most secret fastnesses."

Two explorers gained particular notoriety, though not for their contributions to electric illumination.

One was a world traveler named James Ricalton. "I want a man to ransack all the tropical jungles of the East to find a better fiber for my lamps . . . in the palm or bamboo family," Edison told him. Ricalton agreed, and dutifully searched the forests of Indochina, Ceylon, and India. A year to the day later, he returned to the same New Jersey dock to be greeted by a horde of reporters and the inventor himself, who seized his hand and cried, "Did you *get* it?" Well, no, he hadn't, but his safe return made another good story for the wizard, who hadn't bothered to tell Ricalton he already had a steady supply of perfectly good bamboo.

Then there was Frank McGowan, a mining prospector whom Edison sent to the wildest place in the world, the upper Amazon jungle. With reporters scribbling (at a distance) about his every move, McGowan braved poisoned arrows, infernal reptiles, and the river itself. After fifteen months, he returned to the United States—like Ricalton, without the great filament, but much celebrated for his bravery. On the evening of his arrival in Manhattan, McGowan was toasted at Mouquin's Café. He left the restaurant, wandered off toward the West Side, and was never seen again. His disappearance generated as much copy as his Amazon adventure, but neither police nor reporters turned up a trace of him.

By the time "God's workshop" had been scouted thoroughly, Edison had tested some six thousand organic fibers from all over the globe. None worked as well as an artificial cellulose developed by the English chemist Joseph Swan, who had forsaken adventure for "a little theory and calculation" in the lab. Tungsten, which Edison had considered but found too tricky to use with the tools available at Menlo Park, became the preferred filament in the early 1900s.

A replica of Edison's first operational incandescent lamp. Its carbonized cotton filament was made from sewing thread. It glowed for forty hours.

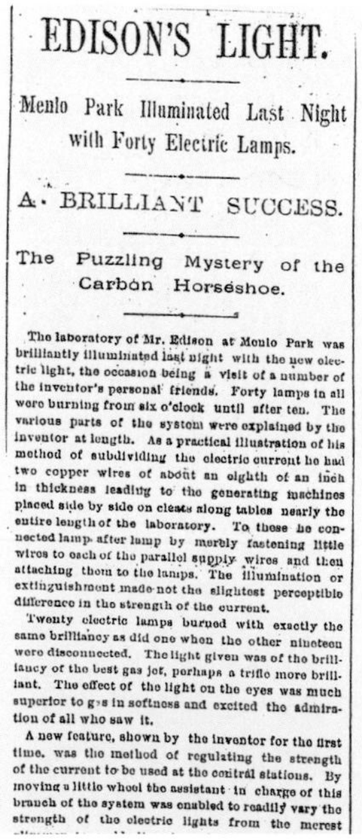

EDISON'S LIGHT.

Menlo Park Illuminated Last Night with Forty Electric Lamps.

A BRILLIANT SUCCESS.

The Puzzling Mystery of the Carbon Horseshoe.

The laboratory of Mr. Edison at Menlo Park was brilliantly illuminated last night with the new electric light, the occasion being a visit of a number of the inventor's personal friends. Forty lamps in all were burning from six o'clock until after ten. The various parts of the system were explained by the inventor at length. As a practical illustration of his method of subdividing the electric current he had two copper wires of about an eighth of an inch in thickness leading to the generating machines placed side by side on cleats along tables nearly the entire length of the laboratory. To these he connected lamp after lamp by merely fastening little wires to each of the parallel supply wires and then attaching them to the lamps. The illumination or extinguishment made not the slightest perceptible difference in the strength of the current.

Twenty electric lamps burned with exactly the same brilliancy as did one when the other nineteen were disconnected. The light given was of the brilliancy of the best gas jet, perhaps a trifle more brilliant. The effect of the light on the eyes was much superior to gas in softness and excited the admiration of all who saw it.

A new feature, shown by the inventor for the first time, was the method of regulating the strength of the current to be used at the central stations. By moving a little wheel the assistant in charge of this branch of the system was enabled to readily vary the strength of the electric lights from the merest

EDISON'S ELECTRIC LIGHT.

What Will Revolutionize the Present Method of Illumination.

CHEAPER AND BETTER THAN GAS.

How the Wizard of Menlo Park Guards His Secret.

The alarm among the English gas companies and the panic in their stocks on account of Edison's last invention in electric lights, as narrated in yesterday's HERALD cable despatches, are not without good cause. America's great inventor has in truth solved the problem which for years has puzzled the ablest electricians of the age. He has successfully divided the electric light, and made it, for illuminating purposes, as far superior to gas as gas is to the the tallow candle of the past. His marvellous invention gives a mild, soft, yet brilliant light, pleasing to the eye, clear, steady and without blemish, and at a cost less than one-third of that required for gas. Those who have seen the invention pronounce it wonderful. Already a company has been formed, composed of a number of wealthy capitalists, and Mr. Edison's lawyers are engaged in completing the final arrangements for the legal transfer. Before long, therefore, the work of introduction will begin. The patents for the United States have only just been granted, while those for the various countries of Europe have not yet been obtained. To-day the Professor's solicitors send by steamer the necessary documents to their London agents for the procurement of the British letters patent. The French and other patents will be applied for simultaneously with the English one. When word is cabled Mr. Edison that the

The *New York Herald* tracked Edison's progress especially closely.

looked at least as daunting as the creation of the electric light itself. To install a working system of electrical lighting over many square blocks, he now would have to invent on an industrial scale. And technology was only half his problem. The other was administration, hardly one of the inventor's strengths. Clearly, he could no longer run the whole show. The great campaign must be divided among lieutenants.

Francis Upton was to stay in Menlo Park, with responsibility for perfecting and manufacturing the large number of lamps that would be needed in New York. Charles Batchelor was dispatched to organize exhibitions of Edison's inventions in Paris and London; the European market was seen as a major source of revenue. The machine shop operation at Menlo Park was transferred to new quarters on Goerck Street in Manhattan.

The job of managing Edison's personal affairs fell to a little British clerk, Samuel Insull, only twenty-one years old. Across the Atlantic, press accounts had led Insull to worship Edison from afar, then to wangle a job with the inventor's London agents, who soon recommended him to the wizard himself. At first, Edison's roustabouts treated this owlish, bespectacled secretary with disdain. But Edison quickly regarded him as indispensable. It was a joining of master and student that would have profound consequences. For his part, Insull recalled in old age, he was immediately "imbued with the idea that I had met one of the master minds of the world. I was young and enthusiastic, and it is true that Edison [had] a peculiar gift of magnetism. But I have never changed my mind from that day to this."

Of course, Edison could not electrify New York without permission of the city fathers. As with his Wall Street financiers, he invited the Board of Aldermen to Menlo Park for their own private exhibition. Well-oiled with refreshments, the pols were soon hollering for a speech from the wizard, and they were delighted to find he was no nose-in-the-air professor. "Why, he looks like a regular fellow," one of them said to another. "See how he handles his cigar? Just like the boys in the [Tammany Hall] Wigwam." The board granted Edison the right to "lay tubes, wires, conductors and insulators, and to erect lamp-posts within the lines of the streets and avenues, parks and public places of the City of New York, for conveying and using electricity or electrical currents for purposes of illumination."

For his headquarters Edison leased a double brownstone mansion at 65 Fifth Avenue. At the roofline he ordered a single word inscribed in block letters—"Edison." But this location was just window dressing. The more important choice of location had begun months earlier, when the inventor assigned aides to prowl Manhattan, eyeing gas facilities, counting customers, poking around in backstreets and alleys. By early 1881 Edison had picked the spot for his First Lighting District—fifty square blocks near the southern tip of the island, the heart of New York's oldest neighborhood. With the locus of residential life drifting uptown, this district was now home to the industries driving American expansion (railroads, mining, oil, telegraphy), the bankers and lawyers who served them, the city's great newspapers, and grimy warehouses and shops. Canvassers went door to door, asking questions. How many gas globes in the building? How much gas is used for heating? For lighting? Have you had gas leaks? How much gas do you use in summer, and how much in winter? The survey yielded exhaustive data, plus a foothold for selling the power that Edison promised would be not only better, but cheaper.

For his central station, Edison bought side-by-side storefront warehouses, 4 stories high and 100 feet deep, at 255–257 Pearl Street. Pearl wound north from Battery Park three blocks from the East River. It was "the worst dilapidated street there was," according to Edison, who raged at the $155,000 price tag for his new properties. (With typical overoptimism, he had hoped the whole project would cost no more than that.)

Towering structures made a rough ring around the district. To the north loomed the great western tower of Washington Roebling's Brooklyn Bridge, nearly complete. To the west were two of the world's tallest buildings, the 230-foot Western Union building, where telegraph operators on the brilliantly lit eighth floor worked 24 hours a day, and the 260-foot Tribune Tower. Looking east, one saw a forest of masts rising from the slips of South Street. Once a place of fashionable residences, it was now a mixture of labor, trade, and high finance, with Wall Street itself for its southern boundary. The smell of the Fulton Fish Market, three blocks from the central station, floated in the air. Various tough customers mixed in the streets—seamen, toughs, immigrant day laborers, cops, newspapermen, and capitalists.

"The Pearl Street station was the biggest and most responsible thing I had ever undertaken," Edison said later. "It was a gigantic

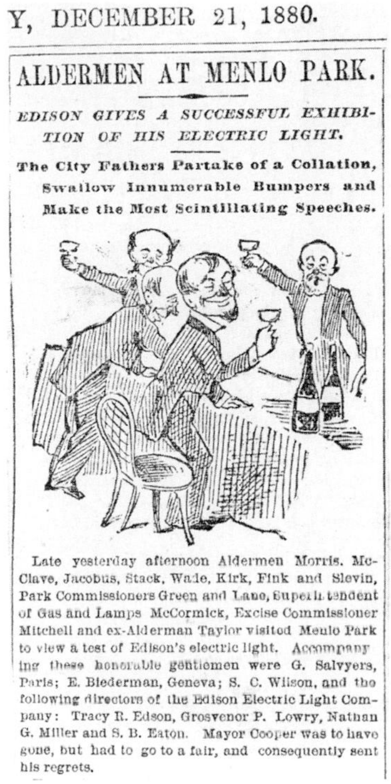

Y, DECEMBER 21, 1880.

ALDERMEN AT MENLO PARK.

EDISON GIVES A SUCCESSFUL EXHIBITION OF HIS ELECTRIC LIGHT.

The City Fathers Partake of a Collation, Swallow Innumerable Bumpers and Make the Most Scintillating Speeches.

Late yesterday afternoon Aldermen Morris. McClave, Jacobus, Stack, Wade, Kirk, Fink and Slevin, Park Commissioners Green and Lane, Superintendent of Gas and Lamps McCormick, Excise Commissioner Mitchell and ex-Alderman Taylor visited Menlo Park to view a test of Edison's electric light. Accompanying these honorable gentlemen were G. Salvyers, Paris; E. Biederman, Geneva; S. C. Wilson, and the following directors of the Edison Electric Light Company: Tracy R. Edson, Grosvenor P. Lowry, Nathan G. Miller and S. B. Eaton. Mayor Cooper was to have gone, but had to go to a fair, and consequently sent his regrets.

The *New York Truth* of December 21, 1880, describes the Menlo Park meeting between the New York City Board of Aldermen and Edison to discuss his plans to light the city. After being plied with a sumptuous meal catered by Delmonico's Restaurant and charmed by a decidedly down-to-earth Edison, the aldermen assented to his wishes to bury the electrical mains underground, which, Edison believed, would be safer than having the wires overhead.

The staff of the Edison Lamp Company, c. 1882.

problem. . . . There was no parallel in the world. . . . What might happen on turning a big current into the conductors under the streets of New York no one could say. . . . Success meant world-wide adoption of our central-station plan. Failure meant loss of money and prestige and setting back of our enterprise."

The Pearl Street warehouses had been built for lightweight commerce. So Edison ripped out the wooden guts of 257 Pearl and replaced them with structural iron. (The old wooden floors were good enough next door, where Number 255 was used for storage, bunks, offices, and labs.) On the top floor, the new girders had to bear the burden of six dynamos, the largest electrical apparatuses ever built up to that time. Edison's crew called them "Jumbos," after the great elephant P. T. Barnum had just brought to the United States—though at thirty tons each they considerably outweighed Barnum's new star. The dynamos were infested with mechanical "bugs." They

Edison had two bases of operation during his quest to light New York. The first was the central station at Pearl Street. The other was this business office on Fifth Avenue.

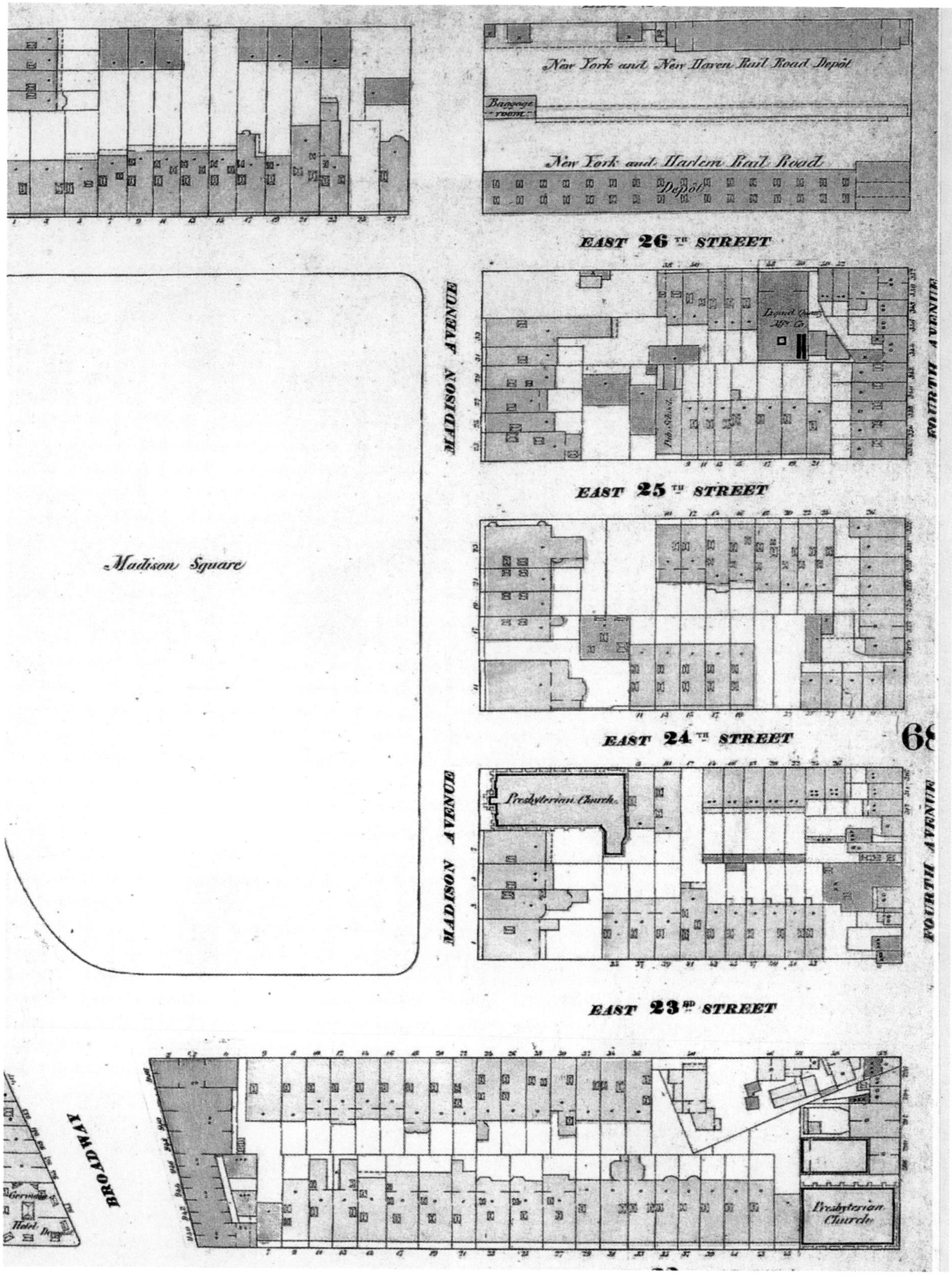

Insurance map of New York City, 1853. To determine the borders of the First District he would light, Edison looked at insurance maps such as this one, which depicts an area north of his First District.

Edison's First District in lower Manhattan. Edison chose this one-square-mile area, bounded by Nassau Street, South Street, Wall Street, and Peck Slip, as the test site for his plans to light New York. The First District was strategically located near the city's financial and newspaper centers. The two lines crossing the East River indicate the almost-completed Brooklyn Bridge.

demanded three months of fine-tuning before they worked reliably, and even then they ran so hot they often had to be packed in ice. Their armatures spun at 325 r.p.m. on the steam power of Babcock and Wilcox boilers in the basement, which consumed eight tons of coal a day.

In Victorian Manhattan, towering utility poles were draped with mile upon mile of cables serving private telegraphs, telephones, stock tickers, and burglar alarms. Edison wanted none of that for his electric lines. He was determined to bury them, like gas lines—a safer but more expensive approach than stringing the wires overhead. Planting them underground demanded excellent insulation, both to protect the conductors from invading moisture and to protect passersby on the streets overhead from escaping electrical charges.

Edison set up a little shop dedicated to insulation on Washington Street. (The storefront was so small the pipes stuck out the windows.) The mains themselves were made of solid copper bars twenty feet long, their cross sections shaped like a half-moon. One half-moon bar was for outgoing current, the other for returning current; they were put face to face with insulation between. To prevent the

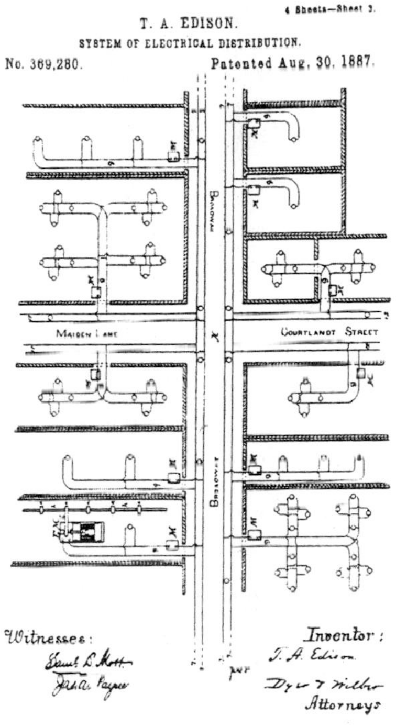

Drawing from Edison's application for a patent for his system of distributing electricity, 1880.

Jumbo Number 9, the first dynamo to provide electric power in New York.

A Pearl Street dynamo.

current from leaking out, workers wrapped these twenty-foot sections in hemp, threaded the sections through iron tubing, then pumped the tubes full of a mixture of Trinidad asphaltum, linseed oil, paraffin and beeswax. The stench of this sticky brew gagged the neighbors, but it was so effective and durable that many of these conduits carried current for half a century.

Meanwhile, Edison hired seventy "wire runners"—men experienced with telegraph lines—to wire up customers' homes and offices. For inside wiring, the copper was wrapped in cotton soaked in shellac, paraffin, or resin, then encased in cardboard sheaths.

In December 1881, Edison was ready to put his mains in place. The city decreed the work must be done between 8 o'clock in the evening and 4 o'clock in the morning, so Edison sent out each of his work gangs with a wagon equipped with a dynamo and an arc light to illuminate the work site. In winter cold and summer heat, ten men would start off tearing up a strip of cobblestones next to the sidewalks. Ten more followed, digging a trench. When the trench was ready, more men set the heavy iron pipes in place, connected them, and ran smaller lines to the buildings. They spurred each other with jests about bringing the gas companies low; in the Edison ranks, an unreliable man was said to "lie like a gas meter."

Edison himself was living in two worlds. If he spent a day at the Fifth Avenue office, he might entertain J. P. Morgan, or the great actress Sarah Bernhardt, or the press magnate Joseph Pulitzer. Often he

The dynamo room at Pearl Street, the first electric light station in the world.

Francis Upton worked with Edison to develop this 1881 dynamo. Its conversion of energy from steam power to electric was far more efficient than that of its rivals.

Jumbo Number 9—the Dynamo Which Supplied the First Current to the Public

A French publication's depiction of Edison's Pearl Street Central station, 1882. On the first two floors, coal was burned in boilers to heat water to create steam in the pipes behind the burners. The steam rose through the pipes to the top floor to drive the dynamos.

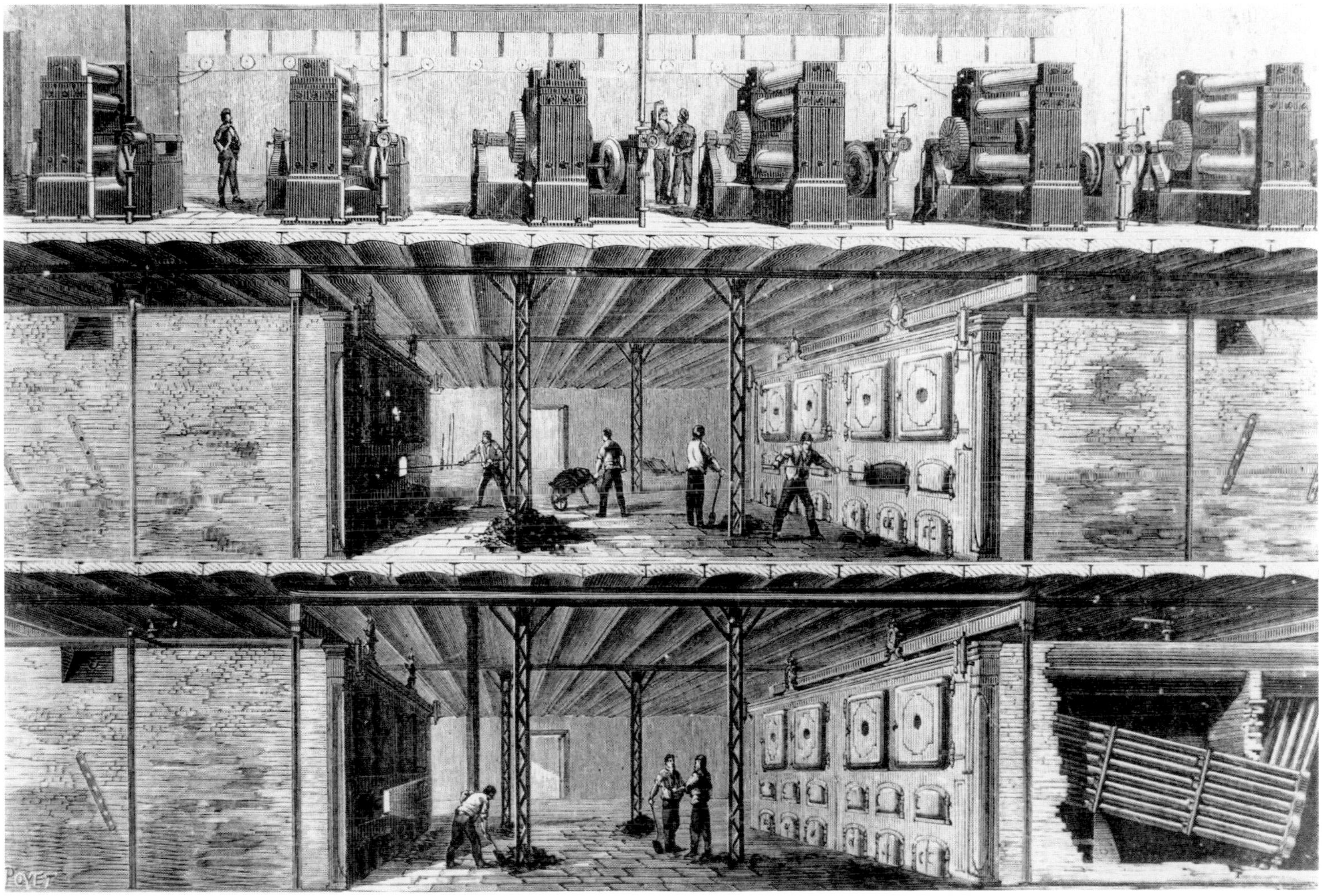

Wall Street during the blizzard of 1888, New York City. Because the city was already overwhelmed with wires, Edison believed it would be safer to bury his underground. Wires shown here were used for the telegraph, the telephone, and arc lights.

took a midnight "lunch" uptown at Delmonico's, his favorite restaurant. But afterward he would go back to Pearl Street, clamber down into the trenches, look over the men's shoulders, issue orders, wrestle with iron fixtures in the cold dirt. His wife and children were ensconced in a hotel on fashionable Gramercy Park, but, despite Mary's deepening depressions and illnesses, he was seeing less of them than ever. Pearl Street absorbed all his energy.

His crews had some fifteen miles of underground lines to lay, and the pace was slow. As the work dragged into the spring and summer of 1882, the Edison crews simply became part of the neighborhood scene. Once a week, well-dressed minor Democrats—appointed as safety "inspectors" by the city machine at Tammany Hall—dropped by to collect their "fees," then wandered off without a second glance. Virtually the only excitement came when Edison began to test the lines in late summer. At the corner of Nassau and Ann, where there was a puddle on the pavement, any horse that started to pull a coach or wagon through the water started and reared in alarm. The animals' shoes were found to be conducting electicity leaking from an underground main. Edison quickly plugged the leak.

Mains were the circulatory system of Edison's creature, but he needed countless other pieces and parts, too, most of which had to

Edison hired about seventy former telegraph "wire runners" to lay more than 80,000 feet of conductors undergound. They worked from the spring of 1881 to the summer of 1882.

The New York Edison Company's runners.

THE ELECTRIC CHAIR

The war of the currents—direct versus alternating—was waged in part with a novel weapon: the electric chair.

Edison learned about electricity in the telegraph business, which relied on batteries and the only kind of current a battery can make: direct current (DC), which flows around a circuit in one direction only. When the inventor began to use generators, which create electricity by passing coiled wire through a magnetic field, he stuck with DC. But DC had a problem: It could be transmitted for little more than a mile.

The electrical engineer Nikola Tesla, who once worked for Edison, and the inventor George Westinghouse, a brilliant newcomer to the field, solved that problem by developing alternating current (AC), which pulses back and forth through its conductor and can be transmitted for many miles. Westinghouse marketed the first AC system in 1886, posing a serious threat to Edison's preeminence. But AC had a problem, too. It required much higher voltage than DC. When AC entered a living organism, animal or human, the result was generally disastrous.

When fully developed, AC would rule the electric future. But in the mid-1880s, Edison concluded it was too dangerous. He always had insisted electricity was safer than gas. Now he insisted DC was the safer of the currents. It was also *his* current, in his own mind and everyone else's, which perhaps made him fonder of DC than was wise. In any case, he was eager for any chance to one-up Westinghouse.

In the fall of 1887, Edison received a letter from a Buffalo dentist named A. P. Southwick, whom the New York legislature had asked to find a humane alternative to execution by hanging. Southwick, who had seen a man accidentally (and instantly) electrocuted, asked Edison for his opinion, saying "science and civilization demand some more humane method than the rope." Edison said he would prefer "an effort to totally abolish capital punishment," but he began to perceive an advantage in Southwick's campaign: If AC became the state's official weapon against murderers, that could hardly help Westinghouse's campaign to sell it for everyday use in the home. So he advised Southwick to look into AC generators, "manufactured principally in this country by Geo. Westinghouse."

When a New York engineer named Harold Brown sought support for experiments in electrocution, Edison lent Brown the use of his West Orange, New Jersey, lab and the assistance of a top aide. Night after night, Brown killed cats, dogs, and finally a calf and a horse. (Animal rights activists were not put out. On the contrary, the Society for the Prevention of Cruelty to Animals asked Edison to help them develop an electric alternative to drowning unwanted animals.) To make his point quite clear, Edison proposed that the new process be called "Westinghousing." (He conveniently ignored the fact that Westinghouse reduced the voltage of AC, and thus the danger, before it ever reached the consumer.)

The state of New York approved the new method of execution, and in 1890 a convicted murderer named William Kemmler became the first man to be "Westinghoused." The process was botched, and Kemmler died in "an awful spectacle, far worse than hanging." Said Westinghouse himself, disgusted: "They could have done it better with an axe."

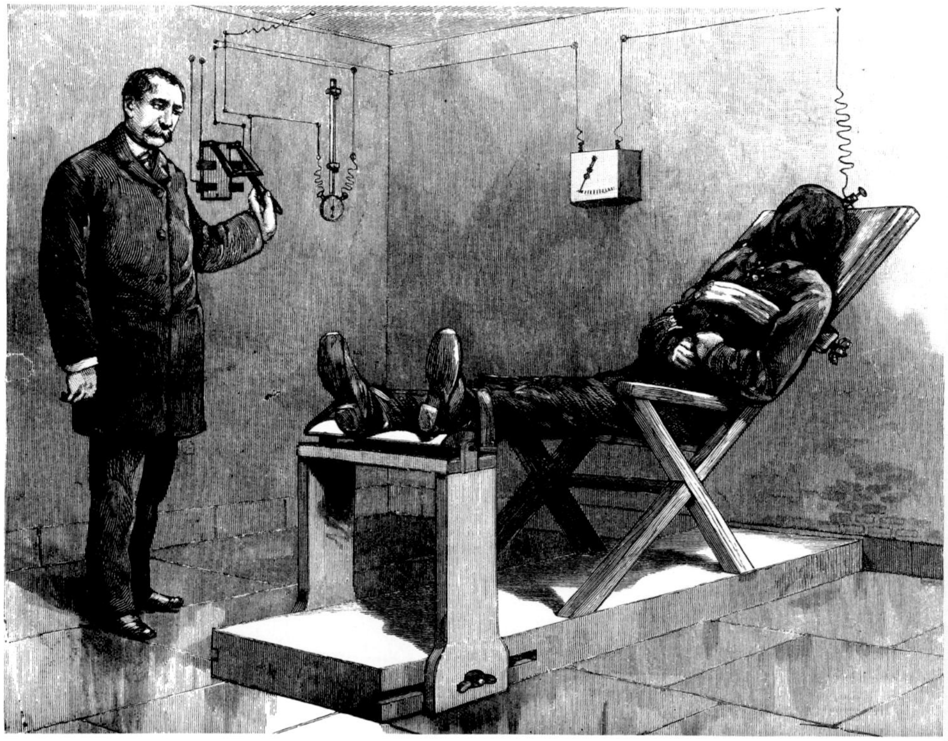

Edison at 34, in 1881.

A wire runner, 1882.

be conceived out of thin air. To manufacture these things, Edison recruited an old associate from his telegraph days in Newark, a German-born mechanic, Sigmund Bergmann. Edison provided Bergmann capital for an expanded factory at Seventeenth Street and Avenue B, on the Lower East Side. Soon, along with lamps made in Menlo Park and generators made on Goerck Street, Edison incorporated a stream of small parts and appliances made in Bergmann's shops—fuses, sockets, switches, junction boxes, safety catches, insulators, voltage regulators, lamp shades, multi-globe hanging lamps called "electroliers," and dozens of other fixtures. Thus did Bergmann & Co. become the forerunner of the electric products industry.

Many of the new gadgets were Edison's own designs. In fact, he applied for no fewer than 107 patents in 1882, the most of any year in his career. One of them was a meter to measure his customers' use of electricity. In each meter, a fraction of the customer's current

An 1882 cartoon in *Judge* depicts the commotion that ensued when current leaked from an underground wire, shocking a horse.

caused drops of zinc to accumulate on a plate. When a meter man weighed the plate, he could calculate the total current used since the last weighing.

Salesmen scouting the neighborhood reported that many potential customers were balking. One reason was the high cost of the lamps: a dollar apiece, a half-day's wages for most workingmen. So Edison offered a dozen lamps free to each customer who signed up for service, inaugurating a free-bulb policy that continued in the United States for decades. To get light, the user turned a thumbscrew at the base of the lamp, which had "the shape of a dropping tear, broad at the bottom, narrow in the neck," and emitting a light that was "soft, mellow and grateful to the eye." It would be many years before this marvelous object became generally known as a "lightbulb."

Running the multiple dynamos in harmony was Edison's greatest worry. On a Sunday in July 1882, with the neighborhood devoid of its weekday bustle, Edison decided to try just two of them together. The first started smoothly, as usual. But when he connected the second, he said later, "of all the circuses since Adam was born, we had the worst then!" A shower of sparks, then flames and a thunderous racket, sent assistants racing for the doors. The whole iron framework of the station's interior writhed up and down. Edison and one other man kept their cool and shut the machines down.

The problem lay with the governing mechanisms on the steam engines down below. Edison called for new designs, but there wasn't time to install them. As usual, Edison explained, "I kept promising through the newspapers that the [central station] would be started at such and such a time. These promises were made more with a view to keeping up the courage of my stockholders, who naturally wanted to get rich faster than the nature of things permitted." Or perhaps the stockholders simply felt that Edison at some point ought to meet a deadline. By summer's end, he could put them off no longer, and he decided to put a part of the district on line—though with just one dynamo.

He chose Monday, September 4, 1882. The night before, Edison kept men up until all hours, checking and rechecking the system's components and rehearsing their assigned roles. "If I ever did any thinking in my life it was on that day," he said.

Edison put on a Prince Albert coat, a white cravat, and a tall, white derby hat. Shortly before three in the afternoon, he and John

Lieb, his chief electrician, pulled their watches from their pockets and synchronized them. Lieb stayed in the station as Edison, John Kruesi, Sigmund Bergmann, and several others strode out the door, turned left on Pearl, then right on Wall Street, and walked two blocks west to the offices of Drexel, Morgan & Co. and up the stairs to the chambers of J. P. Morgan himself.

At 3 P.M. Lieb, at Pearl Street, threw the master switch. In the Drexel, Morgan offices, Edison himself completed the circuit in 106 lamps. Fifty-two more lamps came on in the offices of the *New York Times,* and more still in stores along Fulton and Nassau streets.

Outside, the bright sun of a summer afternoon reduced the moment to anticlimax. But by evening people noticed what had happened: Edison's lights were on.

The *New York Herald,* which had covered the story closely from the start, reported the next morning that "in stores and business places throughout the lower quarters of the city there was a strange glow last night. The dim flicker of gas, often subdued and debilitated by grim and uncleanly globes, was supplanted by a steady glare, bright and mellow, which illuminated interiors and shone through windows fixed and unwavering.

"From the outer darkness these points of light looked like drops of flame suspended from jets and ready to fall at every moment. Many scurrying by in preoccupation of the moment failed to see

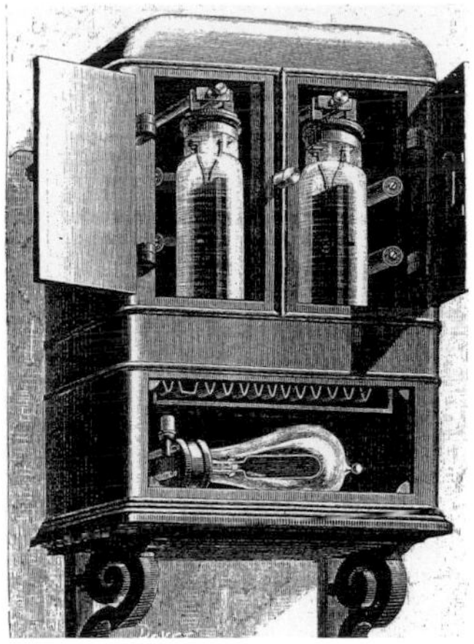

Edison's chemical meter for measuring electricity use, 1882. As early as 1878, Edison was sketching instruments to measure the electricity used in lighted establishments. At that time, he proposed electrifying a silver or copper cell and weighing the deposit from the cell to determine the amount of electricity used. In his First District meters he used zinc plates.

Philadelphia, c. 1883. A Philadelphia Edison Electric Light Company's salesman sets out to sign up new customers for "the most perfect light in the world."

The older Edison. Francis Upton recalled: "His greatness was always clearly to be seen when difficulties arose. They always made him cheerful, and started him thinking; and very soon would come a line of suggestions which would not end until the difficulty was met, or found insurmountable."

them, but the attention of those who chanced to glance that way was at once arrested. It was the glowing incandescent lamps of Edison, used last evening for the first time in the practical illumination of the first of the districts into which the city had been divided.

"The lighting, which this time was less an experiment than the regular inauguration of the work, was eminently satisfactory. Albeit there had been doubters at home and abroad who showed a disposition to scoff at the work of the Wizard of Menlo Park and insinuate that the practical application of his invention would fall short of what was expected of it, the test was fairly stood and the luminous horseshoes did their work well."

Later that day, back at the Pearl Street station, Edison told reporters, "I have accomplished all I promised." In fact he had accomplished both less and more. He had said he would light 2,500 buildings in New York with a single circuit, even 10,000, even an "infinite" number; in fact, he had lit only 400 lamps for 85 customers. But that wasn't the point. These lamps, throwing light across a few rooms and out a few windows, demonstrated the practicality of a system that would transform the world. This success allowed Edison to push on, promoting his vision of central stations as a universal source of electric light and power. The era we recognize as our own, in which energy flows to machines in every American place, began on that summer night in New York.

But the changes wrought by electricity would be a long time coming. In the Pearl Street district, service expanded by small steps. All the dynamos came on line, and by the following fall some eighty-five hundred lamps were in use. But electricity remained an expensive commodity, produced by wholly private concerns and available only to the comfortable classes. Edison had produced the technological means by which electricity could be supplied to everyone, but not the social will to do so. In New York, well-to-do streets were wired over the next few years, but there the wiring stopped. There and in other cities, the gaslight companies struggled on, thanks to electricity's slow advance. Elsewhere, most people still got along with kerosene. Night in Manhattan became a strange patchwork of light, wrote William Dean Howells, "the moony sheen of the electrics mixing with the reddish points and blots of gas."

CHAPTER FOUR

ELECTRIC NATION

Look from a distance at night, and it is as if the earth and sky were transformed by the immeasurable wands of colossal magicians.

—*Murat Halstead at the World's Columbian Exposition, 1893*

On a spring evening in 1892, Samuel Insull arrived at Delmonico's, the most elegant restaurant in New York, to find fifty settings of fine china and crystal awaiting him under sparkling electroliers. Also waiting were forty-nine well-tailored men, the most important figures in the American electrical industry. They had come to honor him as he set off on his new venture in the West—*him,* Sammy Insull, the little London paper-pusher, barely ten years off the boat from Liverpool.

After the men dined on French fare and wine—no wine for the guest of honor, who had promised his mother never to touch it—Insull listened to his hosts pay him tribute. Some were true friends, and some, he knew, were not. When, as an awkward youngster barely out of his teens, he had crossed the Atlantic to become Thomas Edison's personal secretary, some had scorned and teased him. Other men at tonight's table had been his and Edison's adversaries in the battle to light New York. Some were haughty financiers whom Insull had begged for money to keep the chaotic Edison empire afloat. Here, too, was the wizard himself, buoyant as ever at forty-five, the brilliant big brother who had awed his young aide, inspired him, coached him, and eventually relied on him to clean up the messes a worshipful public never saw.

At thirty-three, Insull could look back on an enormous leap in professional standing, of which this dinner was a glittering symbol. He came from a clan of outsiders—lower-middle-class dissenting Protestants, disdained by the proper Anglicans of Victorian London. His father was a ne'er-do-well temperance crusader. To escape a caged future of humdrum jobs, Insull at fourteen went to work by

Samuel Insull in his mid-thirties in 1894, about a year after he arrived in Chicago to take the reins of Chicago Edison.

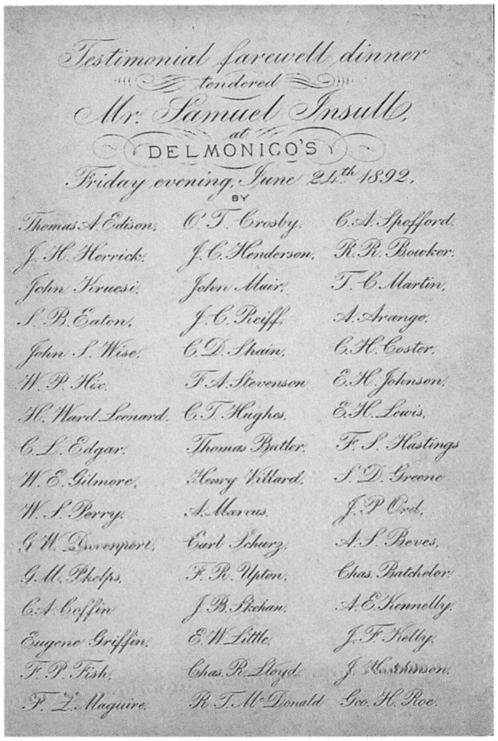

Testimonial farewell dinner
tendered
Mr. Samuel Insull,
at
DELMONICO'S
Friday evening, June 24th 1892.
BY

Thomas A. Edison,	C. T. Crosby.	C. A. Spofford.
J. H. Herrick.	J. C. Henderson,	R. R. Bowker.
John Kruesi.	John Muir.	T. C. Martin,
S. B. Eaton,	J. C. Reiff.	A. Arango.
John S. Wise.	C. D. Shain,	C. H. Coster.
W. P. Hix.	F. A. Stevenson	E. H. Johnson,
H. Ward Leonard.	C. T. Hughes,	E. H. Lewis,
C. L. Edgar.	Thomas Butler.	F. S. Hastings
W. E. Gilmore,	Henry Villard.	S. D. Greene
W. S. Perry.	A. Marcus.	J. P. Ord,
G. W. [illegible]	Carl Schurz.	A. S. Beves,
G. M. Phelps,	F. R. Upton,	Chas. Batchelor.
C. A. Coffin	J. B. Skehan.	A. E. Kennelly.
Eugene Griffin.	E. W. Little.	J. F. Kelly.
F. P. Fish,	Chas. R. Lloyd.	J. [illegible]
F. L. Maguire.	R. T. McDonald	Geo. H. Roe.

The guest list for Insull's farewell dinner in New York, including "most of my intimate friends and intimate enemies."

day and schooled himself by night, embracing the rags-to-riches faith in upward mobility through self-discipline and diligence. These virtues brought him to the attention of Edison's men in England, then served him well in the United States, where he became indispensable to the great man. With fierce energy and a brilliance of his own, a brilliance of organization and efficiency, he began his life's work—to transform Edison's vision of an electrified nation into reality.

Insull's early career was a vertical blur. Starting as Edison's personal secretary, he quickly became his key troubleshooter. After five years in that role, he moved the inventor's manufacturing works to Schenectady, increased the staff from 200 to 6,000, and multiplied the profits many times. ("We never made a dollar until we got the factory a hundred and eighty miles away from Mr. Edison," Insull said later.) When J. P. Morgan engineered the merger of Edison's interests with a competitor, the Thomson-Houston Electric Company, to form General Electric, Insull agreed to become second vice president in charge of all manufacturing and sales. But he made it clear he wouldn't stay long.

G.E. was moving toward a specialization in electrical equipment and appliances; Insull, the carrier of Edison's dream, wanted to run central stations like Pearl Street, only bigger and better. Let others fabricate tools; he wanted to make and distribute power. And he was fed up with the Morgan men behind G.E., whom he regarded as another bunch of big shots holding a leash on the smarter, harder working Sammy Insull. So he sought the presidency of faraway Chicago Edison. It meant a cut in salary from the princely sum of $36,000 a

End of the work day, Edison Machine Works, Schenectady, New York, 1886. When Edison started the factory that year, he put Insull in charge.

year to $12,000. But in Chicago Insull could prove himself on his own, and he had plenty still to prove. If any immigrant kid of the 1880s had made it in America, he had. Yet even a Delmonico's banquet failed to make the outsider feel welcome on the inside.

As one man after another stood to toast Insull, several made wry remarks about his decision to abandon New York for the provinces. In electricity circles, many were saying that "it seemed a pity that Insull was taking up a situation which offered him so little scope as that of the Chicago light and power business." Though Chicago numbered more than a million residents, the New Yorkers still regarded it as a cow town, a slaughterhouse of a city. Chicago Edison was just one of thirty electric companies there and it had only five thousand

Interior, Edison Machine Works, 1886. Insull's first workforce consisted of two hundred men. Six years later, when he decided to leave for Chicago, he had been so successful that six thousand were employed at the factory.

customers. Nobody thought that number would ever exceed 25,000.

Finally it was Insull's turn to speak. He rose and thanked his hosts. Then he put them on notice. In time, he promised, little Chicago Edison would exceed the worth of General Electric. Nearly half a century later, Insull still remembered with pleasure that "this remark of mine caused a great deal of amusement."

Certainly Chicago needed light. Chronically overcast, the city in winter needed help to see by 4 o'clock in the afternoon. Squinting crowds made their way through fogs of soot from the coal-fired boilers that heated nearly every building in the downtown "Loop," and the gloom of the urban canyons seemed to infect the people. Arriving in Chicago about this time from his native Wisconsin, the young architect Frank Lloyd Wright saw "crowds . . . intent on seeing nothing. . . . The mysterious dark of the river with dim masts, hulks and funnels hung with lights half-smothered in gloom reflected in black beneath. I stopped to see, holding myself close against the iron rail to avoid the blind, hurrying by."

North Branch of Chicago River,
looking North from Division Street Bridge, Chicago, Ill.

North branch of Chicago River, looking north from Division Street Bridge, c. 1895. Considering a move to Chicago, Insull worried that his refined tastes might rebel when exposed to the city's industrial wastes. He negotiated a long-term contract with Chicago Edison partly to discourage any impulse he might feel to flee to New York or London.

Yet for the decade before Insull arrived, the business of lighting Chicago was in chaos. The Loop was a battlefield of competing factions and systems: gaslight versus electric; arc lamps versus incandescents; overhead cables versus underground trenches; central stations like Pearl Street versus isolated plants in big facilities such as theaters and hotels, some of which sold their excess capacity to their neighbors. Well-capitalized companies vied with fly-by-nights.

Only the rich could afford electricity. It was the energy of the swanky department stores, the glamorous restaurants, the bosses' office suites. To celebrate his silver wedding anniversary, the tea mogul John Doane ordered 250 lamps installed at his Prairie Avenue mansion; the cost was $7,500, more than ten times what many factory workers made in a year. Of course, electricity's air of wealth made it more desirable. And anyone could see the light itself was superior to gas—brighter, cleaner, safer, prettier. But as long as it cost a penny to light one lamp for one hour, electric current would flow only to the elite, and the rest of Chicago would fight the night with the same old sooty gas lamps. That made it all the more poignant when, just as Insull was getting to work, Chicago staged the World's Columbian Ex-

The Administration Building and the Grand Basin, Columbian Exposition, Chicago, 1893. Insull viewed the World's Fair as an opportunity to demonstrate the promise of electricity.

The electricity building in the Columbian Exposition. Note the spotlight beaming from the tower at right.

position of 1893, a stunning display of what the world might be like if electricity belonged to everyone.

In a century of great exhibitions, this World's Fair was considered the greatest. Sprawling along the Lake Michigan waterfront, it was conceived as a "dream city," a three-dimensional window into a future of astonishing technological achievement. Chicagoans called it the White City, for its monumental classical architecture and its 90,000 incandescent lamps. Electric generators provided the fair not just with its energy but with its message. This was not the bestial force of a steam engine. It was elegant alchemy that changed whatever it touched. "Look from a distance at night," the journalist Murat Halstead wrote in *Cosmopolitan,* "and it is as if the earth and sky were transformed by the immeasurable wands of colossal magicians." Some 22 million fair goers toured the fairgrounds on quiet, fast electric trains and sleek electric gondolas. They stared at a 70-foot "Tower of Light" made of 10,000 colored Edison lamps, "an electric fountain. . . . petrified, as it were, in the midst of its play."

This might as well be Utopia.

Without knowing it, Chicago now had a man who yearned to redeem the promise of the White City—to provide electricity to all the people of Chicago, rich and poor—and who possessed the skills to do so. "Whatever you do, Sammy," the Wizard had once told him, "make either a brilliant success of it or a brilliant failure. Just do something." Edison needn't have worried. Inside this small, nose-to-the-grindstone clerk there was a Napoleon. His mind was not that of a scientist. It was a mind that naturally grasped the nature of technological systems—harmonious arrangements of machines, money, and people working in harness to produce and distribute commodities. Enjoyment matched aptitude; he loved to organize. Edison invented in his lab, Insull at his desk, using arcane charts and graphs. Alfred Tate, who succeeded Insull as Edison's private secretary, said of Insull: "His devotion to business almost constituted a religion. He permitted nothing to interfere with his duties towards the interests he was handling. . . . He loved power and glorified in the exercise of authority."

Yet in 1892, Insull faced such enormous obstacles that it was no wonder his "intimate friends and intimate enemies" at Delmonico's that night in New York had doubted his ambitions in Chicago would ever come to much. From his storefront generating plant on Adams Street, Insull looked out at a ring of busy rivals. In the Loop, his chief competitor was the Chicago Arc Light and Power Company (CALPC), much larger than Chicago Edison and better connected in the city's business elite. Elsewhere in the city, no fewer than sixteen other central-station companies were delivering power to pockets of affluence, and some five hundred businesses were lighting their premises with their own self-contained electric systems; that is, their own on-site generators powering their own lamps and machines. This didn't include the gas companies, which still lit most of Chicago.

All the central-station operators were ensnared in the same economic conundrum. A central station cost a hell of a lot of money. To cover these high capital costs, their owners needed all the big customers they could get—factories, department stores, and office buildings. Yet the economics were such that the more electricity a big customer used, the more money he would save by unplugging from the central station and installing his own self-contained system. If all the big customers did that, the central station would be left with only the home owners and small businessmen, who didn't use enough electricity to cover the central station's massive capital costs.

Then there were technical barriers. Even if Chicago Edison could get all the big customers, it would be hard-pressed to deliver the power they needed. The company's generators on Adams Street already were strained to the point of self-destruction. The station "suggested a glimpse of Dante's Inferno," one employee said later. "The engines were constantly pushed to their utmost capacity under the heavy overload. In the roaring dynamo room, the smell of shellac and varnish from the hot armatures told the same story. Switches and bus bars were frequently too hot for the bare hand, while in the boiler room the half-naked firemen, shoveling coal with feverish energy, made one feel as if an explosion might furnish a climax at any moment." As for delivering current to the bulk of small businesses and average homeowners, that seemed a distant prospect at best. Electricity was just too expensive.

C. Norman Fay, president of Chicago Arc Light and Power Company, the city's biggest electric company in 1892, was more a gas man than an electricity man, and more a high-stakes promoter than either. As head of the city's gas trust, which one historian has called "the perfect epitome of corporate arrogance," Fay was planning a similar consolidation in electricity. With this in mind, he invited the newcomer in town, young Samuel Insull, to lunch at the posh Chicago Club. No doubt Fay had seen the construction crews at an empty railroad site on Harrison Street, where Insull was said to be planning a new generating plant. But Fay wasn't worried. His company was twice the size of Insull's.

Over their meal, Fay told the youngster he was seeking partners to underwrite a major stock issue at CALPC. Would Insull care to join? The young Napoleon said no, thank you. In fact, he said, CALPC, in his opinion, was financially unsound, and it soon would fail, whereupon he, Insull, would buy it. Fay chuckled. If anything, he told his guest, matters would go the other way. Within a year, C. Norman Fay was out of a job and Chicago Edison owned CALPC's franchise and equipment. When Insull's huge new generators at the Harrison Street plant—the world's largest by far—came on-line in 1894, they sent current not only to Chicago Edison's own growing base of customers but to all the old CALPC customers as well.

Harrison Street was by far the world's largest power plant. Insull had assigned the task of designing it to two first-rate engineers—Frederick Sargent, who had worked for Insull during the Edison days, and Louis Ferguson, a young Chicagoan. Sargent and Ferguson equipped the great station with condensing generators, which burned only half as much coal as older plants, and they made sure the plant could be expanded later to meet growing demand. They also installed new equipment that resolved the long-standing "battle of the currents."

Frederick Sargent was one of Chicago Edison's leading engineers. With Insull, he and others at the company pioneered every major advance in power production for a quarter of a century.

Direct current, or DC, was identified with Edison. Alternating current, or AC, was identified with Edison's rival, George Westinghouse. Both currents had advantages; both had weaknesses. DC, which flows around a circuit in one constant direction, moved at a low enough voltage, or pressure, to be safe for use in the home. But DC's low voltage meant the current petered out within a mile or so of the generator. AC, developed by Westinghouse and Nikola Tesla, constantly reversed direction in the circuit at a rate of roughly 16,000 times per second. AC could be sent along the conductor at much higher voltage than DC, so it went much farther than DC. But the voltage was so high that AC was considered unsafe for domestic use.

Insull's man Louis Ferguson looked squarely at the strengths of both systems and proposed to put them in harness together. He bought two new devices called rotary converters. One he attached to a DC generator at the Harrison Street plant; the other he set up in a substation several miles away. The first converter changed DC to AC, which traveled over the wires at high voltage to the second converter at the substation. The second converter changed the AC back to DC, reduced the voltage, and sent it along the remaining short path to consumers. The current could be converted to whatever voltage the customer needed—high voltage for the street railways, for instance, or low voltage for a woman ironing clothes.

With this breakthrough, Insull's physical plant and market penetration expanded hand in hand across Chicago. By 1898 he had bought up every other central-station system in the Loop, and Harrison Street doubled its output. Yet even this amount of generating capacity proved insufficient for the enormous demand for current that arose when Insull took over Chicago's electric streetcars and elevated railways. He had studied the rail companies and concluded he could supply them with electricity at a lower price than what they

The Curtis Steam Turbo-Generator, "one of fourteen 10,000 horsepower electrical units which are being installed in the Commonwealth Electric Company's new Fisk Street Station." One such generator can still be seen on the lawn of General Electric headquarters in Schenectady, New York.

Fisk Street Station. To meet the needs of his growing customer base, Insull realized that "what was needed was the production of energy on a very large scale at a very low cost." The result, the Fisk Street Station, would revolutionize electricity generation. When it opened in October 1903, it was the largest producer of electricity in the world.

paid to make it themselves. For a time, the electric railways would buy two-thirds of Insull's power, greatly solidifying his financial base. But the Harrison Street generators, less than ten years old, couldn't handle the demand, and no one knew how to increase their capacity. They were powered by reciprocating steam engines, whose mechanisms heave back and forth in a pounding, jarring rhythm. Bigger reciprocating engines than Harrison Street's would pound themselves to pieces and might take the building down with them.

When technical limitations loomed, Insull simply pressed his engineers to innovate. After all, he had watched Edison at work. On a trip home to England, he had seen a new kind of engine powering a speedboat—a turbine, or rotary, engine. Its works didn't pound up and down or back and forth; they spun around in a smooth, rotary motion. Insull asked his engineers if they could build a turbine engine for a power station. They shrugged; it had never been done. When he said it was time to try, they balked. So Insull went to the engineers of General Electric in New York. As he expected, they, too, said a turbine engine of the size Insull wanted could not be built, confirming one of Insull's favorite lines—that G.E. engineers could prove anything was impossible.

Next Insull met with Charles Coffin, president of G.E. and one of Insull's hosts at the Delmonico's farewell dinner of a decade earlier. If G.E. men couldn't make the turbine, Insull mused, he knew

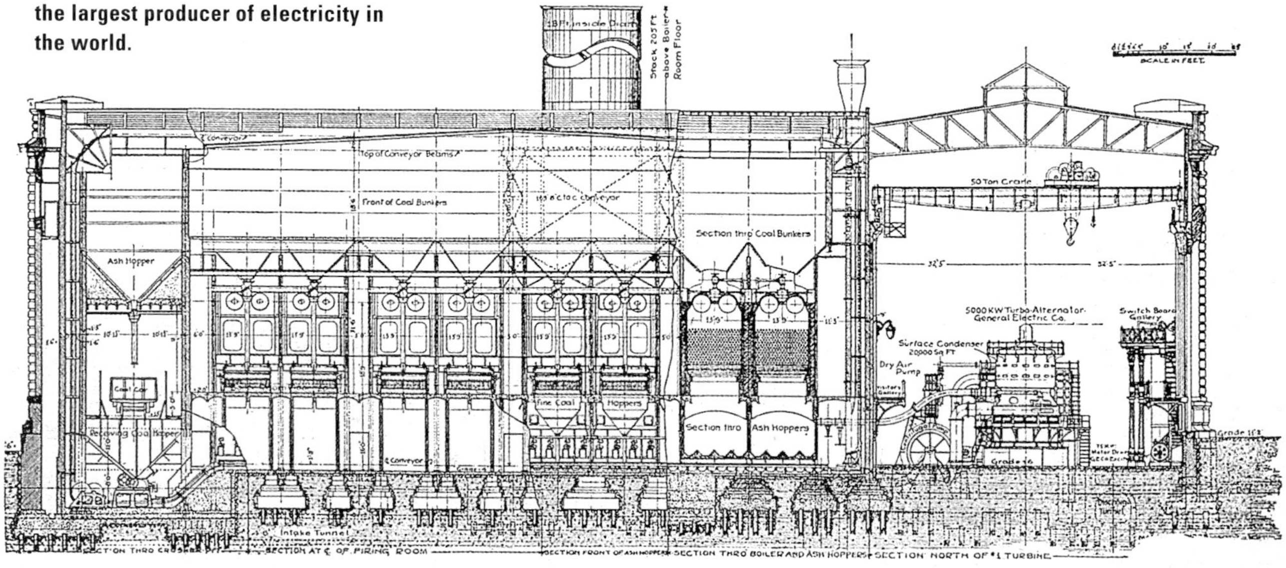

FIG. 6. Cross-Section, Fisk St. Plant, Chicago (*Western Electrician*).

a man in England who could. With that humiliation in prospect, Coffin reluctantly agreed to try. G.E. engineers surprised even themselves and, in 1902, the new engines were installed in Chicago Edison's new Fisk Street Station—once again, the world's largest.

On the day of the plant's dedication, Frederick Sargent, Insull's chief engineer, was about to fire up the turbines when he warned his boss to step away. Insull asked why.

"This is a dangerous business," the engineer replied.

"Then why don't *you* leave?" Insull asked.

"Look, Mr. Insull," Sargent said, "it's my job to stay here. I have to. But you don't. Don't you understand? This damned thing might blow up."

"Well," Insull said, "if it blows up, I blow up with it anyway. I'll stay."

The turbine engines remained intact, though they did have to be replaced by engines twice as big a year and a half later, just to meet new demand.

Three years after that, in 1907, Insull merged his two principal companies, Chicago Edison and Commonwealth Electric Company, to form Commonwealth Edison. It was universally acknowledged as the world's leading utility company. Insull now stood as the sole supplier of electric power to America's second largest city.

Insull marked the twenty-fifth anniversary of the Fisk Street Station in 1928.

As Insull increased his capacity to generate electricity, he was devoting himself even more zealously to the business of generating more customers. For he was the first to perceive that the electric future depended as much on the arcane minutiae of marketing and selling as it did on giant generators and ingenious transformers. As the years passed, the ingenuity and aggressiveness of his marketing reshaped city and countryside and changed what Americans meant by "the pursuit of happiness."

Not long after arriving in Chicago, Insull had told his salesmen to sell at all costs—to drop prices however low they must to make a sale. This had brought phenomenal growth. But they were selling light in the dark. For several years, Insull, like all electric utility operators, struggled with the question of how much to charge his customers.

The difficulty arose from the nature of the product. Electric power is different from other commodities. Winter coats, for example, can be manufactured in summer and stored in a warehouse until the weather turns cold and people need them. The manufacturer doesn't have to produce all the coats he will sell at once, using a huge supply of machinery and labor; he spreads the production out over time. The maker of electricity doesn't have that luxury. His product is manufactured, transported, sold, and consumed all in one moment. He must have enough equipment on hand to meet the peak demand; if he doesn't, he risks brownouts and blackouts. For a good part of every day, much of his expensive equipment will stand idle. But he has to have it in order to meet those times of peak demand.

To pay the high fixed costs of this large capital investment, utility operators went after the biggest customers—factories, office buildings, street lighting, electric railways. To win those big contracts, they cut their rates competitively, but rates stayed high for the home owner or small business, whose business wasn't attractive. As long as rates remained high, consumption was stuck far behind the consumption of gas.

Insull glimpsed the possibility of a new approach in his native land. In 1894, he took his Christmas holiday in England. (Britain remained Insull's official home until 1896, when he became a naturalized U.S. citizen.) Feeling poorly in the London fogs, the young batchelor boarded a train for Brighton, the pretty beach resort on the island's southern coast. There he knocked about by himself, curious to see electricity at work among his compatriots, who had embraced municipal ownership of utilities. Insull was startled to see the bright glow of electric lights burning all evening in every shop window. In Chicago, the high cost of current prevented small customers from keeping their lights on for so long. Curious, Insull sought out the young man in charge of Brighton's central station, Arthur Wright, who told Insull about the electric meter he had invented. Until now, meters had simply measured the total number of kilowatt hours a customer used in a month, and the customer paid a flat rate to match. Wright's meter showed not just how much current the customer used, but when he used it—how much during peak hours, how much during off hours.

At home, Insull thought this over, then sent Louis Ferguson back to Brighton to study Wright's methods in-depth. With Ferguson's report in hand, Insull sat at his desk, pondering charts and scribbling scenarios. An idea emerged. He would issue each customer a two-part bill. The first charge would be for the customer's share of peak demand; with this charge he would pay for his share of financing the company's plants and equipment. Naturally, the big users would bear most of this heavy cost, but Insull would woo them with discounts for heavy use. The second charge—much lower—would be for the kilowatts the customer used during off-peak hours. In short, the more electricity a customer used, the cheaper the electricity would be. This deceptively simple notion lit the fuse that electrified the nation.

Soon Insull was spending most of his time studying and rewriting his rate schedules. "The way you sell the current has more bearing on . . . cost and profit than whether you have the alternating or direct-current system, or a more or less economical steam plant," he declared. He invited Arthur Wright to Chicago to spread the message. "It is possible," Wright told an industry conference, "that more profit may be derived from the supply of electricity to small, long hour consumers at low rates than from the supply of large consumers, such as extensive stores, at a much higher rate."

Insull used a simple chart to show how Chicagoans used electricity from 12:01 A.M. until the following midnight. The chart showed a line running over peaks and valleys. Naturally, the lowest valley appeared from 2 A.M. to 5 A.M. From 6 A.M. to 9 A.M., the line rose as people awoke, turned on lights, and boarded streetcars and elevated trains for their jobs in offices, stores, and factories. Another valley followed at noon, as machines were shut down for lunch. Then the line soared to a peak of usage between 4 and 8 P.M., as people rode the streetcars home and turned on their lights, while others worked on in late shifts.

Referring to this chart, the historian Thomas Hughes explained Insull's great innovation this way: "Insull and his associates . . . did everything possible to fill the valleys." Insull perceived that an empire lay at his feet if only he could get people to spread their use of electricity more evenly around the clock. The more they did so, the more he could lower rates, meaning electricity would spread farther and farther down the social scale until nearly everyone could afford it.

Insull now embarked on a permanent campaign to encourage mass consumption of electricity around the clock. He gave deals to all-night restaurants. He encouraged factory owners to put on overnight shifts. He urged the owners of office buildings to install new electric elevators. To increase usage during prepeak hours in the morning and afternoon, Insull sent his salesmen out to talk with housewives. Their pitch wasn't just electric service but electric appli-

Insull created a chain of retail stores around Chicago to sell early electrical appliances. These included the electric baby rocker, the marshmallow toaster, and electric ovens.

ances—washing machines, vacuum cleaners, refrigerators, fans, water heaters, ovens, hot plates, teapots, coffeepots, toasters, curling irons, heating pads. Insull sent trucks loaded with irons up and down city streets, offering free use of an iron for six months to anyone who signed up for central-station service. In the Railway Exchange Building on Michigan Avenue he opened an Electric Shop and stocked it full of the new electric devices, which he promoted in blizzards of company-sponsored circulars and advertisements. To entice factory owners, the shop included an Industrial Power Room loaded with electric lathes, drills, and punches.

In 1898, the first full year of the Wright meter's career in Chicago, the average Chicago Edison customer saw his electric bill drop 32 percent, and that was just the beginning. Insull cut rates over and over and over. The two-tier electric bill sounded the death knell for self-contained electric systems and even for gaslight itself. It was the instrument of what became became known as "load man agement," and it led to a second electrical revolution, as profound in its way as Edison's.

But the revolution was a long time coming, especially in the home. Even in spite of Insull's rate-cutting, only one in six Chicago families used electricity in 1910. The typical family with service used five to seven 50-watt light bulbs (each with about one-third the brightness of today's 50-watt bulb) for two hours in the evening, usually in the dining room and the parlor. In the rest of the home, and during the rest of the day, they made do with gas. Many regarded their electric lamps like fine china—appropriate for company, but not for everyday. Of the other home appliances on the market, the electric iron was by far the most popular, relieving women of the backbreaking chore of heating heavy irons on a stove, then hauling them back and forth over the family's clothes and linens. Yet even here, women used the electric irons only sparingly, saving them for summer heat, when the work was the most onerous.

Out in the streets the electric tide was far more obvious to an observer. The cool, steady glare was illuminating the life that everyone shared, and changing it. The new lamps glowed in stores, offices, penny arcades, and midways. They lit the screens of the new five-cent movie houses, where teenagers swarmed every night to gape at Edison's latest idea. "What is seen there becomes the sole topic of conversation," huffed the great social worker Jane Addams, "forming the ground pattern of their social life."

An advertisement in *Woman's Work*, c. 1900.

Libertyville, Illinois, 1910. Insull's purchase of an estate outside the small town northwest of Chicago was the first step toward an electrical empire that would reach through the Midwest and beyond.

Chicago's suburbs: Highland Park, Evanston, Waukegan, and so on. The suburbs were linked into a single grid, and the suburban grid was linked to the city's grid. From northern Illinois he marched to Indiana, then throughout the Midwest and beyond, forming major companies as he went. By the mid-1920s, Insull's networks were supplying current in thirty-two states, and his customers used one-eighth of all the electricity and gas consumed in the United States.

The electric future Edison had prophesied in Pearl Street now came to pass in the cities of the 1920s, with Chicago leading all. By 1925 virtually every home in the city used electricity, and most families owned at least an electric iron and vacuum cleaner, too. The radio, coming to market in 1923, fast became the most popular appliance of the lot—"the first true mass-consumption electrical product," the historian Harold L. Platt called it. In a statement that could apply to any item in the century's endless parade of technological innovations, one sales document declared that "almost before the householder realizes it, he is relying on electricity for his light and various other needs and wondering how he could have gone without it so long."

Insull presided over the world he had made like a benevolent despot. Under the surface of the twenties, one could hear faint rumblings of unease about the all but unfettered power of the new utilities, but Insull himself commanded immense respect. In an era of prosperity and soaring stock values, with electricity seeming to fulfill every promise of an easier and better life, he was revered, like Henry Ford and Edison himself, as a seer, a miracle worker, a bringer of fire. Early in his career he had dodged the spotlight; as a boy he always had been told that the only people who got their pictures in the paper were crooks and royalty. But now, perhaps in mute acknowledgment that he had joined the royal species American style, he allowed himself to be lionized. Reporters sought and printed his advice. (Question: "Mr. Insull, what are the most important elements in success?" Answer: "Health, honesty, hard work, firmness of character, firmness of action, imagination, and keeping everlastingly at it.") Professional societies, universities, and governments stood in line to give him honors. As dusk fell, he would stand at the window of his office and watch Chicago's electric lights come on.

During a trip to London in 1926, Insull received a summons from Stanley Baldwin, the prime minister of Great Britain. Insull hurried to 10 Downing Street, where he listened as Baldwin de-

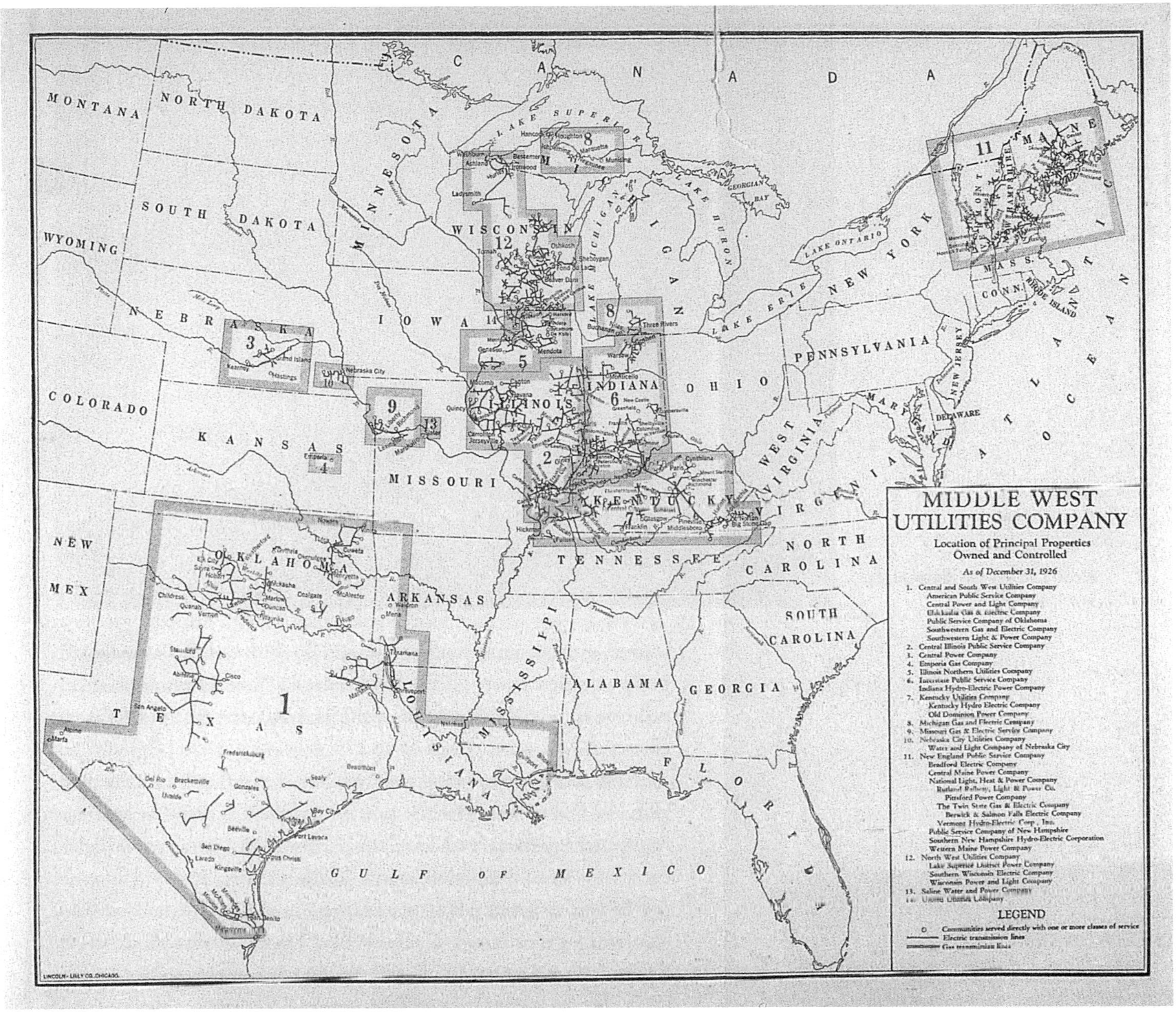

Middle West Utilities Company's holdings c. 1925.

panies to be controlled by himself, his son, and his brother, Martin.

No sooner had this been done than the market entered its giddy final frenzy. Stock in the Insull companies soared. During fifty days in the summer of 1929, the value of Insull stocks rose at the rate of seven thousand dollars per minute. Alternately puzzled, confident, and panicky, Insull and his aides wove an ever more complex web of ownership structures and refinancings. Parsing the Insull holdings later, General Electric president Owen Young, who had helped to reorganize the finances of Europe at the end of World War I, said: "I confess to a feeling of helplessness as I begin to examine . . . the complicated structure of that organization." Unaware of these dangerous interdependencies, awed by the Insull name, half a million people sank all or part of their savings into his securities.

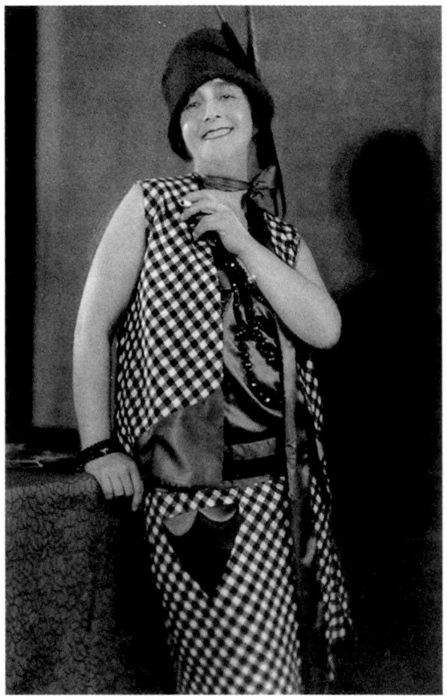

Insull's wife, Gladys, weighed less than ninety pounds. A successful actress, she had left the stage when she married. But in 1925, at the age of fifty-six, she returned to the stage to play an eighteen-year-old ingenue in Richard Brinsley Sheridan's *School for Scandal.* She drew rave reviews.

It took three years for the Great Depression to undo it all. Insull fought mightily if not always wisely. Believing he could single-handedly save his own companies and stave off the general decline, he overextended mightily, taking on enormous debt. He threw all his own money into the fight and took out personal loans. But the pyramid of companies could not withstand the pressure of falling prices. He went hat in hand to the Morgan bankers he had always spurned and asked for a $50 million loan in short-term notes. He got it. But when he missed an installment, the bankers closed in. They declared the holding companies insolvent and floated rumors—unfounded, it turned out—that the Insulls had been guilty of indiscretions at best and embezzlement at worst. In the spring of 1932, with the entire country on its knees, one Insull company after another collapsed into receivership. Only his three key operating companies, including Commonwealth Edison itself, remained solvent. Forced to resign some sixty presidencies and directorships, Insull faced the press and said, "Well, gentlemen, here I am, after forty years a man without a job." A few days later, he and his wife departed for a long rest in France.

With the fabric of national life in shreds, the name Insull, a symbol of everything praiseworthy in the twenties, now came to symbolize all the nightmarish forces that had overtaken the economy. A stock that had been among the most trusted in the nation was now worthless, and people demanded a scapegoat. In Chicago, a Republican prosecutor who had been Insull's friend, and who privately told his own son-in-law that "I know he has never taken a dishonest dollar," sought to save his reelection chances by investigating the Insull

INSULL'S OPERA

Sam Insull drew little applause from the monied families of Chicago's Gold Coast when he drove the price of electricity down to where working people could afford it. In fact, they didn't care much for him generally—though they didn't say so where he could hear it. He seemed unawed by the greater age of their money and unenthusiastic about their social rituals. They didn't approve of his favorite charity (doctors in Africa!), and his wife, Gladys, didn't bend the knee to them. But they cared least of all for what he did with the Chicago Civic Opera.

Opera was one of Insull's passions. In the European tradition, he saw it not as a high-falutin society affair but as the art form of ordinary people. As a teenager in London he sometimes had skipped his supper to save the cost of a seat in the upper gallery. In 1922, when Harold and Edith Rockefeller McCormick stepped down as the opera's leading benefactors, he stepped up to replace them—and to remake the local opera scene.

In perfect Insullian fashion, he began by spreading the load, recruiting a larger number of benefactors who would be asked to pay less. He also said it was time to quit deferring to the opera's star, Mary Garden, who liked to sing in French. "There are four hundred thousand German-Americans in Chicago," he said, "and a quarter of a million Italian-Americans, and not a handful of Franco-Americans." Miss Garden departed after one season. Then Insull went about providing the opera with a new home. He did so with ingenuity and a keen instinct for the aristocracy's jugular.

He solicited a design for a monumental office building, forty-two stories high, with the opera's quarters at street level. (In the unmistakable shape of a giant armchair, the building soon became known as "Insull's throne.") He financed it by selling bonds to the Metropolian Life Insurance Company and preferred stock to ordinary opera lovers. The common stock was held in trust for a new agency, the Chicago Music Foundation. When the bonds and preferred stock were retired (with money raised from rental fees charged to the building's tenants), the foundation would own the building and continue to reap the rentals. The scheme would flood the city's opera and other music projects with money indefinitely. Insull leased the first block of offices himself. The only thing wrong with the plan was its timing. The building was completed in the autumn of the stock market crash.

The world's great opera houses included private boxes, overlooking the stage, for the well-to-do. The boxes' real virtue was not the good view, but the fact that people in the cheap seats below could look up to the boxes to see who was really who. As Insull's project went forward, he publicized the exterior design of the building but kept the interior a deep secret. No one was permitted to get a glimpse until opening night, November 4, 1929. When the doors finally opened, guests wearing plain cloth and fur got the same surprise: No elite boxes overlooked the stage. (There were boxes, but way at the back, where no one could see in.) At Insull's operas, all Chicagoans would sit together, and every seat was a good one. The Civic Opera House still stands as one of North America's most beautiful opera houses, host since 1954 to the Lyric Opera of Chicago.

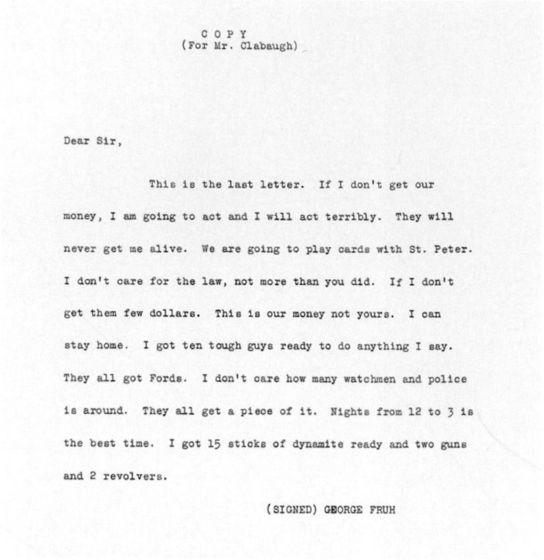

COPY
(For Mr. Clabaugh)

Dear Sir,

This is the last letter. If I don't get our money, I am going to act and I will act terribly. They will never get me alive. We are going to play cards with St. Peter. I don't care for the law, not more than you did. If I don't get them few dollars. This is our money not yours. I can stay home. I got ten tough guys ready to do anything I say. They all got Fords. I don't care how many watchmen and police is around. They all get a piece of it. Nights from 12 to 3 is the best time. I got 15 sticks of dynamite ready and two guns and 2 revolvers.

(SIGNED) GEORGE FRUH

Though many people suffered financial losses as a result of Insull's bankruptcy, his businesses fared better than many during the Great Depression. The average American company forfeited on 40 percent of its securities. For Insull's companies, forfeitures were just over 20 percent.

debacle. At the same time, the Democratic candidate for president, Franklin D. Roosevelt, swore vengeance upon "the power trust" and denounced "the lone wolf, the unethical competitor, the Ishmael or Insull whose hand is against every man's."

In October, just weeks before the national election, Insull was indicted on federal charges that he had promoted and sold securities to investors at inflated prices, thus swindling small investors out of $100 million. Indictments soon followed against Samuel Jr., Martin Insull, and several associates. Insull Sr. had moved from Paris to Athens, since Greece had no extradition treaty with the United States; he said he would return after the election, when politicians would no longer need to crucify him so publicly. After eighteen months of transatlantic legal wrangles, he was hauled back to Chicago under guard and installed in a Cook County jail cell with several lesser known tenants, including a murderer. In defiance, Insull refused bail until the next day.

Reporters swarmed; headlines blared. It was the show trial of a dark time, with Insull standing in for the depression itself, and for all the financial devices of the boom that preceded it. But Insull's lawyer, Floyd Thompson, a former justice of the Illinois supreme court, was fully up to his task, and so was the principal defendant. When the prosecution rested, Thompson put the old man, now sev-

Stock certificate for Chicago District Electric Corporations, one of Insull's companies, 1931, eighteen months after the Black Tuesday New York Stock Exchange collapse. None of the utility magnate's electric and gas operating companies went bankrupt; people who bought stock in them lost only a fraction of a percent on their investment. But investors in his holding companies did lose significant amounts of money.

enty-five, on the stand. Newspapers around the world sent reporters to cover his testimony, and Chicago's business elite jammed the courtroom to stare and listen.

First Thompson drew out Insull's life story—his up-by-his-own-bootstraps childhood in London; his years with the sainted Edison; his central role in the making of modern Chicago. Then Insull fired back at his pursuers point by point. He had formed his later investment companies only to ward off takeover attempts. He had taken not a cent in cash from them. In fact, he had sunk his own savings into them to try to save them and incurred millions in personal debt in the same struggle. He pointed out that his accounting methods, much maligned, were the same ones used by the prosecutors' employer, the federal government. Asked if his annual salaries had come to roughly $500,000, he coolly shot back: "Yes, and I was worth every penny of it. 'The laborer is worthy of his hire.'" Later testimony revealed that Insull's enormous charitable contributions often exceeded his salary.

By the halfway point in his testimony, the shift in the trial's tone had the feds flummoxed. During a recess, Samuel Insull, Jr., found himself standing in the men's room next to Leslie Salter, the lead prosecutor, a New Yorker reputed to be the ablest Justice Department attorney in the country. Looking dazed, Salter murmured to the younger Insull, "Say, you fellows were legitimate businessmen." "That's what we've been trying to tell you," the son replied.

Though the prosecutors had made several Insull associates look pretty bad, Insull made no effort to shift blame to them. "Gentlemen," he said, "if I feel some embarrassment sitting here in this chair, I feel it infinitely greater for these young men who had no responsibility in this situation, and for whose welfare I have almost as much concern as for the defendant who bears my name." He was charged with inflating his securities to make a killing, he said, yet even when the stock ran wild, neither he nor his son nor their associates had sold a bit of it. There had been no cover-up; the prosecution's case had come directly from Insull company records. Insull refused to allow any character witnesses to be called in his behalf, but he declared with ringing pride that no operating company of his had ever failed to meet an

2 CENTS PAY NO MORE!

Chicago Daily Tribune
THE WORLD'S GREATEST NEWSPAPER

HOME EDITION

WEDNESDAY, SEPTEMBER 14, 1932 — 32 PAGES — PRICE TWO CENTS

$226,000,000 INSULL LOSS

MAINE ELECTION RESULT DISMAYS G.O.P. AND DRYS

Democrats, Repeal Forces Jubilant.

NEWS SUMMARY of The Tribune

MAYOR CERMAK LANDS IN N. Y. ON HOMEWARD TRIP

Told Europe Truth About Chicago.

BY TOM PETTEY.

SKETCHES OF THE NEWS

INSULL'S DIZZY DROP FROM $100,000,000 TO 'EXILE' AND PENSION

INVESTMENT CO. AND INVESTORS ARE WIPED OUT

Audit Report Filed in Federal Court.

Flyers Sighted

September 14, 1932. The Insull Utility Investment Company was wiped out after a bank audit found corporate deficits of $226 million.

Insull left jail after posting bond. His legal team had been told that his bail would be set at $100,000, but when he arrived at the Cook County jail, it had been doubled. Insull insisted upon staying in jail for the night rather than appeal to the courts. This was the first move in his effort to win the public relations war that attended his legal battle.

Insull as he awaited a train to return him to Chicago for trial, 1934. Insull became despondent on the trip and reportedly considered suicide until his State Department guard revived his spirits.

Federal prosecutors with piles of government evidence.

Gladys Insull and Samuel, Jr., during the trial.

obligation—never a bill unpaid, a dividend missed, a paycheck dishonored.

The jurors took five minutes to agree on a verdict. When the sheriff told them they'd be accused of taking bribes if they reported so fast, they had cake and coffee for a couple of hours, then declared the defendants not guilty of all charges.

Their decision failed to restore Insull's reputation. If many in Chicago still respected him, many others there and elsewhere could not disentangle his colossal achievements from the bubble of the twenties and the bankruptcies of the Great Depression. He died alone in a Paris subway in 1938, his body unidentified for several hours because a passerby had lifted his wallet. Those who did not hate him soon forgot him, and his name, unlike Edison's, faded. Yet he had done as much as Edison himself, perhaps more, to electrify the nation.

Leslie E. Salter, the most renowned prosecutor in New York, was first called to Chicago to provide advice and moral support to the prosecution. But, on September 14, 1934, the night before the trial, lead prosecutor Forrest Harness was unexpectedly removed, possibly due to exhaustion. Salter was thrust into the role of lead prosecutor.

Norris Dam, c. 1936.

of the TVA) until his death in 1975. Wendell Willkie became the greatest dark horse in American political history, shocking Republican stalwarts by winning the party's presidential nomination in 1940 and nearly denying Franklin Roosevelt a third term in the White House. Lilienthal fulfilled his early promise as head of the TVA, then as director of the postwar Atomic Energy Commission.

Sometimes the workers at TVA dam sites saw an old man standing nearby, watching the earthmovers and the concrete trucks. George Norris liked to watch his old dream take shape.

In the TVA's first months, Lilienthal and Llewellen Evans, the agency's chief electrical engineer, were badly in need of information about how to run a great unified utility system so as to charge customers as little as possible while still providing excellent service. They could not very well ask Wendell Willkie for advice, nor any other utility magnate in the United States. So they crossed the Atlantic to study the government-owned British grid. There they found an excellent model that helped them a good deal in their planning at home. It was left to a small London trade publication, the *Electrical Review,* to remind the two Americans of who had inspired the British grid and given advice on how to build and run it: "Mr. Samuel Insull, of Chicago. . . . Those who know the story of his early and remarkable work in the expansion of electricity production and consumption in the States will remember that it was the envy of students of such matters here. . . . But circumstances have deprived the American people of his experience."

In the verdict of historians, the public and private crusaders for electrification, however much they fought each other, share credit for a monumental work of civilization. Insull's jury acted properly in acquitting him. He had acted not only honestly but with extraordinary vision. But Lilienthal's vision was vindicated, too, in the spread of electricity to places the private utilities might long have bypassed. The electrification of America was an enormous task, too great, perhaps, for either private or public forces alone. They accomplished it together.

Insull, still widely respected, on the cover of *Time* magazine the week after the stock market crash of 1929.

3 METROPOLIS

"The city was more than a spectacle. It was also a great machine for communal life. And it had to work properly if the whole urban enterprise were not to disintegrate."

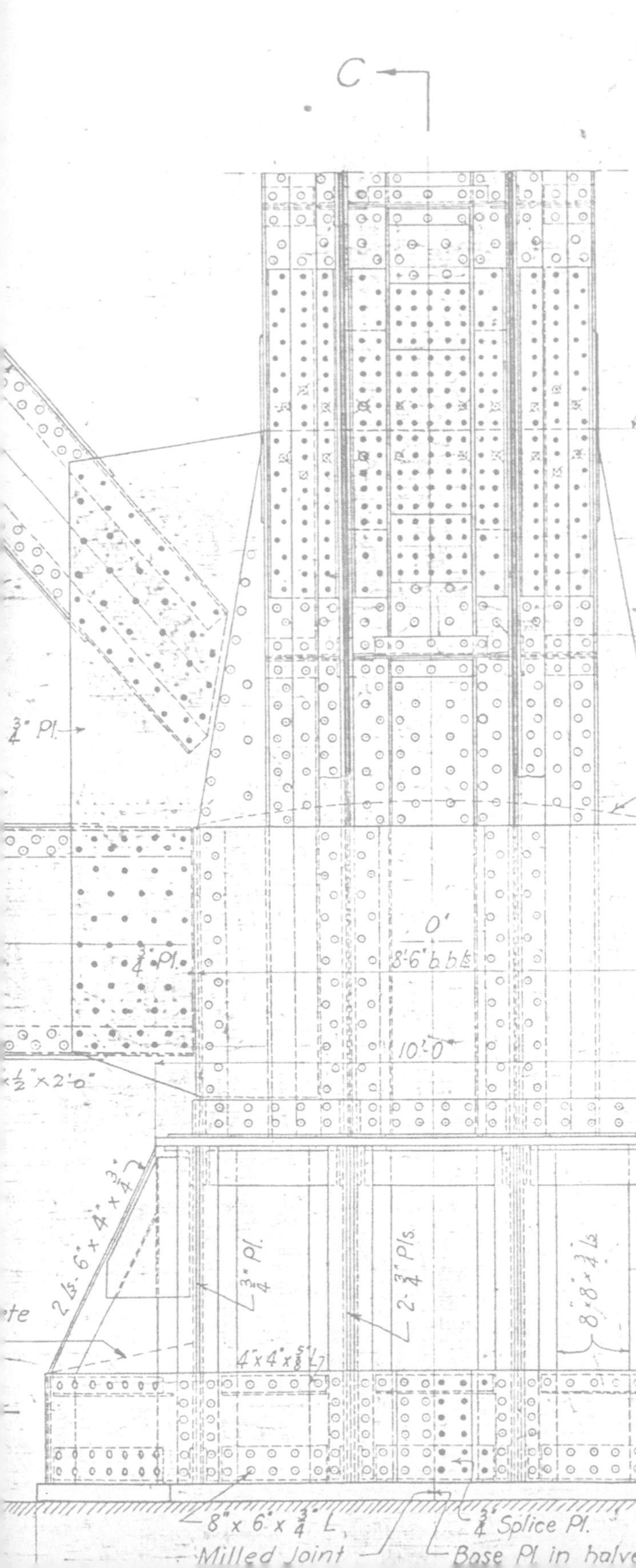

The landscape itself—rivers and mountains—was the first adversary of America's engineers. But in the long run the man-made world of the cities posed challenges at least as great.

In its youth, the United States was principally a nation of farmers. Its few large cities were puny by European standards and unpopular among Americans themselves—"absesses on the human body," said the scientist Benjamin Rush, friend to Thomas Jefferson, who believed the rise of cities threatened American democracy.

But even in Jefferson's lifetime, the seeds of an urban nation were planted. With steam power came factories demanding labor in numbers that only cities could privide, and locomotives and steamboats that linked city to city. From 1870 onward, a chronic farm depression drove plowmen to town in search of work. At the same time, waves of immigrants from Europe cascaded into the Atlantic ports, and after 1910, destitute black farmers fleeing the boll weevil and Jim Crow swelled the cities further. Cities across the country doubled, tripled, quadrupled in population. Most striking of all was the rise of New York, the old colonial port that by World War II would be a world metropolis of some 7.5 million.

Industrial wages drew the newcomers, of course, but not wages alone. In the cities they found wonders never seen in the countryside—Edison's bright lights and the artificial night of the movie houses, gorgeous department stores and thundering rail yards, glamour and noise, romance and mystery. If this self-confident nation's manifest destiny stretched westward to the Pacific, it seemed also to stretch skyward with the great houses of industry and commerce. Here, the energies of labor, finance, art, and science were converging to form a new urban civilization. In reputation as well as size, the unquestioned leader of this new America was New York, with its symbols of urban splendor—the world's tallest buildings and Broadway's "Great White Way," the electric streetcars and the underground trains, and the magnificent Brooklyn Bridge of John and Washington Roebling.

But the city was more than a spectacle. It was also a great

machine for communal life. It was constructed piece by piece over many decades, without a master plan. And it had to work properly if the whole urban enterprise were not to disintegrate.

Of all the components of this new machine, none was more important than a reliable system for providing residents with clean water. Without it, there simply could be no city, at least no great city—a fact that New Yorkers came to realize only gradually and at great cost. In the wake of an epidemic that good water might have prevented, engineers designed a bold solution. It would test the ability of a democratic people, suspicious of state power, to band together to build something that no man or corporation could build alone.

The water system was part of the city's internal mechanisms. Just as important were its connections to the world outside. Indeed, New York rose to power on the strength of those connections—first its ships, plying the routes of world trade; then the Erie Canal, reaching out to the nation's granaries; then the railroads, strongest of all the links between the city and its inland empire.

When the automobile arrived early in the twentieth century, offering Everyman his own means of freedom and speed, New York faced a critical decision. The remaining barrier between city and nation was the broad Hudson River; everyone knew it had to be crossed. But should the crossing be a tunnel or a bridge? And should it carry trains, the symbol of manifest destiny and corporate power in the nineteenth century, or automobiles, symbol of the new century's urge toward maximum individual freedom?

The question fell to two brilliant engineers, one old and one young. They began as mentor and student, then became friends. But the question of the Hudson River crossing made them rivals, and the outcome of their battle would shape the face of the great city.

CHAPTER FIVE

WATER FOR THE MILLIONS

No population but one of freemen would have conceived the idea.

– New York City Water Commission, 1842

In the summer of 1832 everything came to the island city of New York by water—people, animals, goods in trade. There was no other way to reach the city than by ship, boat, or barge. So that was how the Asiatic cholera came, too. Probably it came down the Hudson River from Albany, where the disease had last been seen, in the body of some workman or immigrant or child. The bacillus killed its first victims in Manhattan in the last week of June and took two thousand more by the end of July. Symptoms began with sudden diarrhea, then progressed swiftly to violent nausea, severe muscle pain, and overwhelming thirst. In a day or two the pulse dwindled and the skin turned a faint blue. Robbed of bodily fluids, the victim lapsed into a stupor and died.

Without knowing exactly how cholera spread, New Yorkers had known it was coming, had heard of its progress from India to Europe to the British Isles to Canada. They had read in the newspapers of 20,000 dead in Mecca, 7,600 in Paris, 2,200 in Quebec, 1,800 in Montreal. Those with any means had fled Manhattan in panic by Independence Day, 100,000 strong, half the population. So in the city's comfortable Fifth Ward, where Alderman Myndert Van Schaick kept a lovely home, the streets were still and silent. On the East Side the immigrant slum called Five Points was in shock but hardly silent. Van Schaick, walking the lanes and looking in on the dying, heard moans, weeping, and a single plea repeated many times: "Give me cold water!" But the doctors told Van Schaick there was little clean water to give. By the time the epidemic had receded in the fall, some thirty-five hundred New Yorkers had died. More than any other single event, the cholera of 1832 forced the city to face up to

The neighborhood of Five Points on New York's East Side was home to an immigrant community hard hit by the 1832 cholera epidemic.

its first great technological challenge: It must find a way to supply itself with clean water or it would wither away.

Many perceived the emergency, but it was Myndert Van Schaick who was first to act. He sprang from one of the old Dutch families that settled New York in the early 1600s, then stayed on through English rule as the closest thing in America to a native aristocracy. Born at the end of the revolution, Van Schaick had seen his city grow from a colonial town to a mercantile center poised to become one of the world's great cities. He married the daughter of a family like his own and became a philanthropist and gentleman politician. He was a noblesse oblige Democrat—a wealthy man who believed his class was duty-bound to improve the circumstances of a broad new citizenry of immigrants. To educate those who were unwelcome at snooty Columbia College, he helped found the college that would become New York University. As a city alderman, he became treasurer of the Board of Health, the role that brought him to the cholera wards.

Cholera was a disease of dense, dirty human habitations. Many clergymen attributed it to the moral squalor of slum life. Doctors thought it was caused by "miasmic vapors"—atmospheric particles wafting upward from putrifying organic matter. Some twenty years later, a British physician would prove the bacterium entered its victims through the mouth, in water or food tainted with infected feces. The only way to prevent the disease was to ensure that people drank clean water. Once a victim was afflicted, the cure, too, was clean water, if administered quickly with doses of salt. In 1832 these facts weren't known. But if doctors were ignorant of the scientific details,

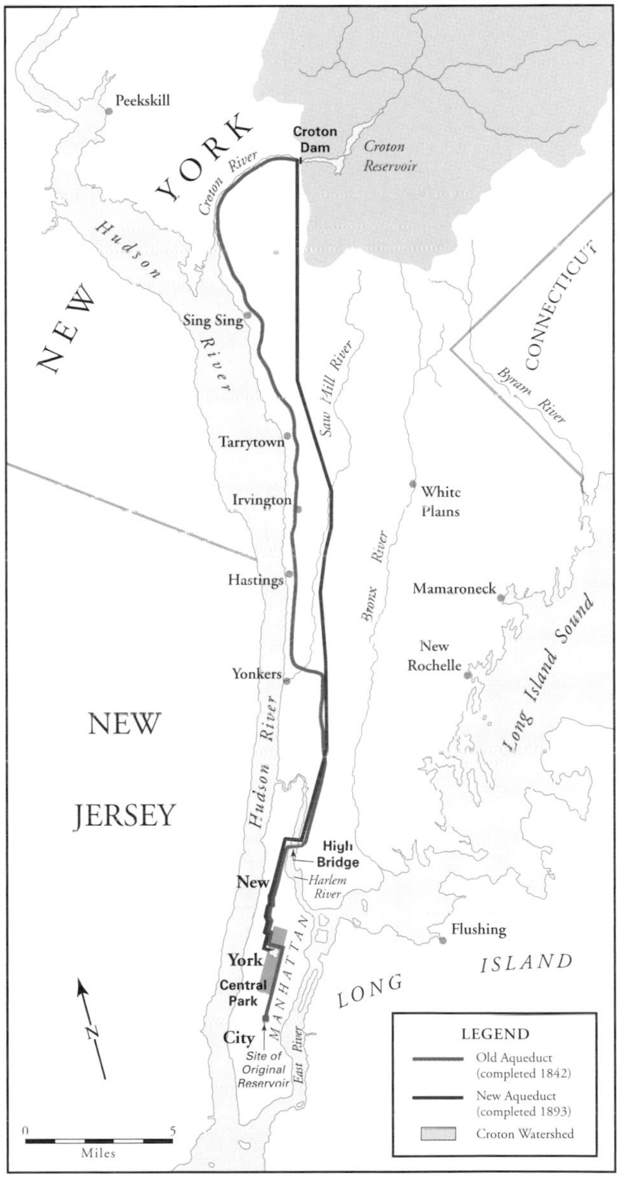

The routes of the first and second Croton aqueducts.

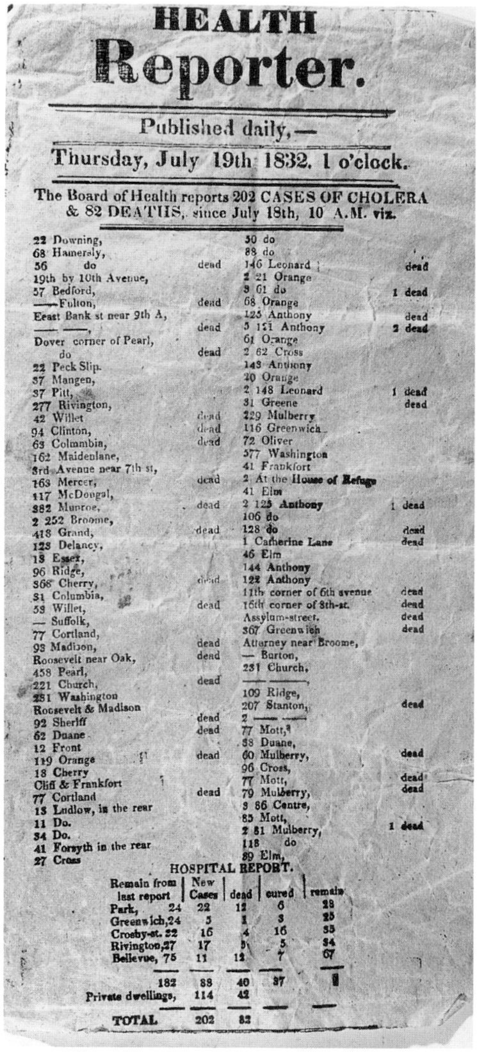

HEALTH
Reporter.

Published daily,—

Thursday, July 19th 1832. 1 o'clock.

The Board of Health reports 202 CASES OF CHOLERA & 82 DEATHS, since July 18th, 10 A.M. viz.

22 Downing,		30 do	
68 Hamersly,		88 do	
56 do	dead	146 Leonard	dead
19th by 10th Avenue,		2 21 Orange	
57 Bedford,		3 61 do	1 dead
——Fulton,	dead	68 Orange	
Eeast Bank st near 9th A,		125 Anthony	dead
—— ——,	dead	5 121 Anthony	2 dead
Dover corner of Pearl,		61 Orange	
do	dead	2 62 Cross	
22 Peck Slip.		143 Anthony	
37 Mangen,		20 Orange	
37 Pitt,		2 148 Leonard	1 dead
277 Rivington,		31 Greene	dead
42 Willet	dead	229 Mulberry	
94 Clinton,	dead	116 Greenwich	
63 Colmmbia,	dead	72 Oliver	
162 Maidenlane,		577 Washington	
3rd Avenue near 7th st,		41 Frankfort	
163 Mercer,	dead	2 At the House of Refuge	
117 McDougal,		41 Elm	
382 Munroe,	dead	2 125 Anthony	1 dead
2 252 Broome,		106 do	
418 Grand,	dead	128 do	dead
123 Delancy,		1 Catherine Lane	dead
13 Essex,		46 Elm	
96 Ridge,		144 Anthony	
368 Cherry,	dead	122 Anthony	
31 Columbia,		11th corner of 6th avenue	dead
53 Willet,	dead	16th corner of 8th-st.	dead
— Suffolk,		Assylum-street.	dead
77 Cortland,		367 Greenwich	dead
93 Madison,	dead	Attorney near Broome,	
Roosevelt near Oak,	dead	— Burton,	
458 Pearl,		231 Church,	
221 Church,	dead	—— ——,	
231 Washington		109 Ridge,	
Roosevelt & Madison		207 Stanton,	dead
92 Sheriff	dead	2 —— ——	
62 Duane	dead	77 Mott,	
12 Front		38 Duane,	
119 Orange	dead	60 Mulberry,	dead
18 Cherry		96 Cross,	
Cliff & Frankfort		77 Mott,	dead
77 Cortland	dead	79 Mulberry,	dead
13 Ludlow, in the rear		3 86 Centre,	
11 Do.		85 Mott,	
34 Do.		2 81 Mulberry,	1 dead
41 Forsyth in the rear		118 do	
27 Cross		89 Elm,	

HOSPITAL REPORT.

	Remain from last report	New Cases	dead	cured	remain
Park,	24	22	12	6	28
Greenwich,	24	5	1	3	25
Crosby-st.	22	16	4	16	35
Rivington,	27	17	5	5	34
Bellevue,	75	11	12	7	67
	182	88	40	37	[illegible]
Private dwellings,		114	42		
TOTAL		202	82		

The *Health Reporter* published the names of those who had contracted cholera or died from it during the previous day. Manhattan's first reported case was that of "Mr. Fitzgerald," an Irish immigrant who became sick on Monday, June 25, 1832. He survived, but his wife and children were dead by Wednesday. Published less than a month after the Fitzgeralds' deaths, this report cites a total for the day of 202 new cases and 82 deaths.

they knew in general what was wrong: New York was overflowing with human and animal excrement.

In his rounds of the poor neighborhoods, where dry layers of aged filth encrusted the streets, Van Schaick heard doctors discuss the desperate need to make the city clean, especially the wards where the poor lived. Indeed, the epidemic may have forced the poor into the everyday consciousness of many well-to-do New Yorkers for the first time. "Only on occasions such as this," said one of the Roosevelts, "is the true extent of the misery of the City known." When the epidemic eased in the fall, Van Schaick went to City Hall and began to read old records on the city's water problem to see what could be done.

". . . THIS ABOMINABLE FLUID . . ."

The Dutch, then the British, chose Manhattan for its seafaring advantages. But the East River and the mouth of the Hudson, so well-suited for oceangoing commerce, were salty mixtures of ocean and fresh water, and thus undrinkable. Colonial New Yorkers drew their water from shallow wells—some private, some public—and a seventy-acre pond where Canal Street runs today. The Dutch called the pond the *kolck* ("small body of water"), which the English bastardized to "collect." For nearly two hundred years, the wells and the Collect quenched the thirst of Manhattanites. But their revulsion at what they were drinking rose in direct proportion to the increase in population.

Early efforts at a collective solution flopped. Just before the revolution, Christopher Colles, a civil engineer, began work on a system of pinewood pipes that would distribute water through much of the city from a new, deep well near the Collect. But the plan was abandoned during the chaos of British occupation. By 1800, when the city's population hit 60,000, one editor said the Collect had become "a shocking hole" into which residents had dumped "all the leakings, scrapings, scourings, p—s—ngs, &—gs, for a great distance around." These included the chemical wastes of several neighboring breweries and tanneries. Nonetheless, people still bathed and did their laundry in the water of the Collect. The "Tea Water Pump," long regarded as the city's last reliable source, lost its prestige precisely because it drew from the Collect. By 1813, the Collect had disappeared under the landfill of builders who built the Five Points slum where the pond had been. But the springs that had nourished the pond still oozed under the ground, spreading the leakings of privies and graveyards from well to well.

Philadelphians had done far better. With the help of the distinguished architect Benjamin Latrobe, designer of the U.S. Capitol building, they were the first to use steam engines to power a water system and the first to use cast-iron pipes, which delivered a "cheap and abundant supply of pure and wholesome water . . . not excelled by any in the world." Of course, Philadelphians enjoyed good geological luck. They were bracketed by the freshwater Delaware and Schulkyll rivers. But their example embarrassed New Yorkers. In Philadelphia, a Manhattan commission reported enviously, "no disagreeable odor assails the persons who pass through the streets . . . but in New York a person coming in the city from the pure air of the country, is compelled to hold his breath."

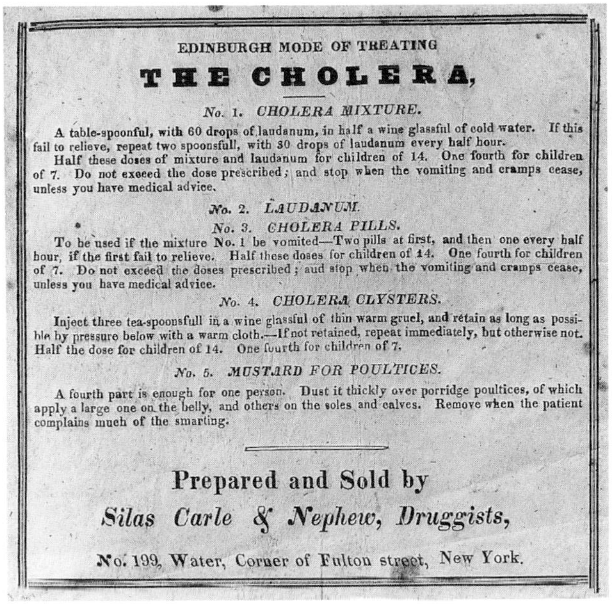

EDINBURGH MODE OF TREATING

THE CHOLERA,

No. 1. CHOLERA MIXTURE.

A table-spoonful, with 60 drops of laudanum, in half a wine glassful of cold water. If this fail to relieve, repeat two spoonsfull, with 30 drops of laudanum every half hour.

Half these doses of mixture and laudanum for children of 14. One fourth for children of 7. Do not exceed the dose prescribed; and stop when the vomiting and cramps cease, unless you have medical advice.

No. 2. LAUDANUM.

No. 3. CHOLERA PILLS.

To be used if the mixture No. 1 be vomited—Two pills at first, and then one every half hour, if the first fail to relieve. Half these doses for children of 14. One fourth for children of 7. Do not exceed the doses prescribed; aud stop when the vomiting and cramps cease, unless you have medical advice.

No. 4. CHOLERA CLYSTERS.

Inject three tea-spoonsfull in a wine glassful of thin warm gruel, and retain as long as possible by pressure below with a warm cloth.—If not retained, repeat immediately, but otherwise not. Half the dose for children of 14. One fourth for children of 7.

No. 5. MUSTARD FOR POULTICES.

A fourth part is enough for one person. Dust it thickly over porridge poultices, of which apply a large one on the belly, and others on the soles and calves. Remove when the patient complains much of the smarting.

Prepared and Sold by

Silas Carle & Nephew, Druggists,

No. 199, Water, Corner of Fulton street, New York.

Cholera remedies included laudanum, a liquid form of opium. It was a pacifier, not a cure.

In the summer and fall of 1798, more than two thousand New Yorkers had died of yellow fever—about one in thirty city residents. A young doctor, Joseph Browne, argued that further epidemics were inevitable unless the city imported a steady supply of water from off the island. His theory, like theories about cholera, was wrong; yellow fever is passed by mosquitoes, not by bad water. But his proposal was eminently healthy.

Browne's idea was to build a system of dams, bridges, and pipes that would bring "remarkably pure and pellucid" water to Manhattan from the Bronx River, which ran from ponds in Westchester County to Long Island Sound near the north end of Manhattan. The system would be built by a new, public-spirited private company, chartered by the state. Water not drunk by New Yorkers and their animals would be made available to wash the streets and fight fires. City officials were enthusiastic.

Then Browne's famous brother-in-law, never one to overlook a chance for self-advancement, stepped in. This was Aaron Burr, just then a member of the New York State Assembly, formerly a U.S. senator and aide to General Washington, soon to become vice president under Thomas Jefferson. Burr and some wealthy friends were eager to establish the city's third bank. But their plan needed the approval of the legislature in Albany, where the existing banks would fight it hard. Burr needed a front. His brother-in-law's popular water plan looked like just the thing.

Through adroit politicking, Burr and his friends turned the Browne plan inside out. When they were done, Burr stood at the

As the city's water supply became more and more contaminated, colonial New Yorkers favored "tea water" from private springs.

Aaron Burr.

head of the state-chartered Manhattan Company. Its stated purpose was to supply the city with "pure and wholesome water." But thanks to a murky little clause that Burr slipped into the charter, the company was authorized to invest in other "monied transactions or operations" in support of the water plan. With this wording, Burr had his bank. For convenience, he abandoned the plan to bring water from the Bronx River and instead resurrected Christopher Colles's abandoned waterworks. Meanwhile, he put his new stockholders' money to work in the Bank of the Manhattan Company.

As a bank, the Manhattan Company did well. In time it became Chase Manhattan. As a water company it was a disaster. It did everything on the cheap. It used wooden logs for pipes instead of cast iron. When the company dug up streets to lay pipes, it left the trenches unfilled, though the inconvenience was minimized by the company's lack of enthusiasm for laying pipes at all. Customers paid high rates that the Manhattan Company set as it pleased. Often the system simply did not work. Those who drank the water regularly blamed it for stomach ailments ranging from the uncomfortable to the lethal. In a letter to the *Evening Journal,* "A Water Drinker" excoriated the

Manhattan Company for "this abominable fluid." True, the writer said, only the desperate had to drink it straight, but everyone had to cook with it: "Our tea and coffee are made of it, our bread is mixed with it, and our meat and vegetables are boiled in it. Our linen happily escapes the contamination of its touch, 'for no two things hold more antipathy' than soap and this vile water."

Of course, the source of contamination was the populace itself, and their horses, hogs, and cows. The animals used the streets. The humans used backyard outhouses built over temporary holding bins. By law, these "privy vaults" were to be made of leakproof stone and to sit at least five feet underground. Godforsaken city workers called "necessary tubmen"—all of them black—were employed to empty the bins and trundle the waste by "night cart" to landfills, fertilizer manufacturers, or the Hudson or East river, to be dumped where the tide would sluice it out to sea. That was the idea. The practice was different. Waste often backed up out of the privy vaults and streamed into the streets, and tubmen sometimes missed the tide, leaving the offal to lap against the island's edges. Many privy vaults were made of leaky wood or brick, not impermeable stone, so their contents sank into the shallow water table and thence to springs and wells. Even Brockholst Livingston, an original director of the Manhattan Company, conceded that many Manhattanites were "drinking a proportion of their own evacuations."

For thirty years the Manhattan Company fought to keep other central suppliers out of the market, just to safeguard its license to run a bank under the shingle of a water company. So New Yorkers cursed the water and got along. The poorest drew the worst water from 250 public wells. Those who could afford it paid twenty dollars a year for water hauled in barrels from cleaner springs at the island's north end. After several hours under a summer sun in dusty streets, the water wasn't very good, but it was better than the stuff from the wells. If you wanted it cold, you had to buy ice, which cost more than the water.

The taste of city water spawned the soda-water industry; in 1828, the price on street corners and at the new "soda fountains" was three cents for a tumbler of artificially carbonated water mixed with a dose of lemon syrup. Those who wanted a stronger draught supported the growing numbers of taverns and breweries. Thus the scarcity of good water indirectly encouraged public intoxication, rowdiness, crime, and vice.

The Manhattan Company's reservoir, shown in this 1825 watercolor, was on Chambers Street, near New York City's present-day City Hall. A bronze Aquarius loomed over the entrance, as if to guard the water supply.

It could not go on. With a population approaching 150,000, New York was twice the size of its closest American rivals, Philadelphia and Baltimore, and the pace of its growth was accelerating. Yet the city lacked an element that was necessary to the survival of even the most primitive human settlement—a dependable source of clean water.

The need was only increasing. As mayor in 1811, DeWitt Clinton had unveiled an ambitious blueprint for expansion—a grid of some 2,000 blocks interlaced by 12 avenues and 155 streets. Sail power connected Manhattan to the ports of every continent, and now, after a quarter century of Napoleonic wars, ships were returning from Europe with crowds of Englishmen, Irishmen, and Germans. Many of them stayed in New York, cramming their families into crumbling two-story structures built for an earlier time. These people had to eat and they had to eliminate. As Edwin G. Burrows and Mike Wallace put it in their authoritative city history, "[N]o matter how well or ill they were made, the city's privies were not prepared for the torrent of shit that now descended on them, courtesy of a proliferating population."

In 1831, a study by the Lyceum of Natural History, the city's leading scientific body, found that New York was staggering under a burden of 100 tons of human and animal excrement per day. City water was found to be dosed with urine and 126 grains of organic waste per gallon, compared to only 2 grains per gallon in the nearest country stream. With dry finality, the scientists declared: "No adequate supply of good or wholesome water can be obtained on this island for the wants of a large and rapidly increasing city like New York."

SO WATER WOULD HAVE TO BE FOUND SOMEWHERE ELSE

Myndert Van Schaick took the lead. His quest, detailed in Gerard T. Koeppel's authoritative *Water for Gotham,* had begun just before the cholera siege of 1832, when Van Schaick joined the Common Council as a "pro-water" man and began to attend meetings of the city's Fire and Water Committee. In the wake of the epidemic, he studied the city's troubled history with water and all the unrealized schemes for importing water from the rivers closest to Manhattan: the Bronx, the Byram, the Saw Mill, the Housatonic. At some point Van Schaick's eye stopped at something unfamiliar—a letter submitted by a defunct canal company asserting that Manhattan could get limitless good water from the distant Croton River, forty miles to the north in upper Westchester County. The idea had been passed over; the distance seemed far too great. But Van Schaick, desperate for fresh possibilities, pursued it.

He enlisted an ally with a helpful political name—DeWitt Clinton, Jr., son of the former mayor and governor, just twenty-seven years old but already a respected civil engineer with experience in river and railroad surveys. He was also well liked. President Andrew Jackson himself called Clinton "a Gentleman hugely respected . . . both for his promising talents & his amiable private character." Clinton sparked to the Croton idea. Soon he and Van Schaick were sketching a daring plan.

As men raised in an avidly democratic culture that revered the Roman Republic, Van Schaick and Clinton took inspiration from the nine stone aqueducts that supplied ancient Rome with clean water. Parts of the aqueducts still stood—great cement-lined pipes that ran underground and over arched bridges, some for more than fifty miles. Beautiful and enduring, they supplied Romans with some 38 million gallons of water each day. No modern city had tried

WATER AND THE MUSE

The taste of Manhattan's water in the years before the Croton Aqueduct was strong enough to call forth the muse. In 1826, Samuel Woodworth, a New York newspaper editor, was moved to dewy-eyed recollections of "The Old Oaken Bucket" from which he had refreshed himself as a lad on his father's farm. Woodworth's elegy was widely republished and became much loved. An excerpt:

That moss-covered bucket I hailed as a treasure,
For often at noon, when returned from the field,
I found it the source of an exquisite pleasure,
The purest and sweetest that nature can yield.
How ardent I seized it, with hands that were glowing,
And quick to the white-pebbled bottom it fell.
Then soon, with the emblem of truth overflowing,
And dripping with coolness, it rose from the well.

How sweet from the green, mossy brim to receive it,
As, poised on the curb, it inclined to my lips!
Not a full, blushing goblet could tempt me to leave it,
Tho' filled with the nectar that Jupiter sips.
And now, far removed from the loved habitation,
The tear of regret will intrusively swell,
As fancy reverts to my father's plantation,
And sighs for the bucket that hung in the well.

Another poet, this one's name lost to history, added these stanzas to Woodworth's:

Oh, had I but realized in time to avoid them
The dangers that lurked in that pestilent draft,
I'd have tested for organic germs and destroyed them
With potassic permanganate ere had I quaffed.

Or perchance I'd have boiled it, and afterwards strained it
Through filters of charcoal and gravel combined;
Or, after distilling, condensed and regained it
In potable form with its filth left behind.

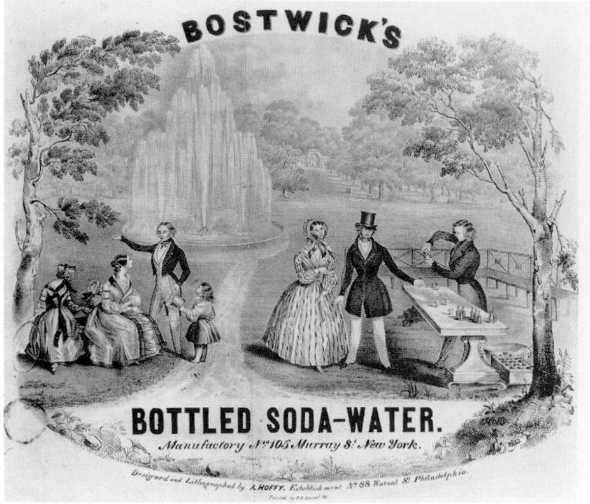

George Usher is said to have first distilled and bottled mineral water in 1811. After his death, the city gave his widow permission to sell his popular product in the foyer of City Hall. Many rival companies, such as Bostwick's, sold competitive products on street corners.

such a project except London, and its aqueduct was just half as long as what Van Schaick and Clinton were contemplating. But if Romans could do it, they reasoned, why not Americans?

Young Clinton's engineering plan for a Croton River aqueduct was sketchy at best, and it offered few details about how water would be distributed once it got to Manhattan. The plan's virtue was its farsightedness. Like his father, Clinton insisted New York would one day number a million souls. Neither the shallow Bronx River nor others close to the city, which others were targeting, could possibly supply so many people. The only practical source for so large a population, he said, whatever the cost, was the deep, cold Croton.

For two years Van Schaick and his allies waged political warfare on behalf of Clinton's vision. They battled well-diggers in the city, property owners in Westchester, and the proprietors of the Manhattan Company, whose right to act as exclusive purveyor of water to the city was slapped down by the state supreme court. Van Schaick felt the people must have the final say, since they would foot the bill for the substantial bond issue. "It seemed necessary to rouse the citizens," he said later, "and to call out a strong sentiment in favor of the measure." So in April 1835, nearly three years after the cholera epidemic, the question of whether to issue bonds for the building of an aqueduct from the Croton River went to the voters of New York.

During a three-week campaign, water dominated talk in the city. Advocates, well aware of public hatred for the Manhattan Company's "odious monopoly," stressed the communitarian nature of the project. "Water is one of the elements," said the newly formed Water Commission, "full as necessary to existence as light and air, and its supply, therefore, ought never be made a subject of trade or speculation." The Committee on Fire and Water agreed emphatically:

> The control of the water of the City should be in the hands of this [municipal] Corporation, or in other words, in the hands of the people. From the wealthy and those who would require the luxury of having it delivered into their houses; and from the men of business, who would employ it in their work shops and factories, the revenue should be derived. But to the poor, and those who would be content to receive it from the hydrants at the corners and on the sidewalks, it should be as free as air, as a means of cleanliness, nourishment and health. In the hands of any other power than the Common Council, this free use would be restrained, and the experience of all other Cities (and our own may be included) teaches us the sad lesson that the trust of this power would be abused.

Democratic newspapers were for it, hastening to remind their working-class readers that their children's health was at stake. Whig newspapers, bullish on "internal improvements" and speaking for and to the business class, were for it, too. On the last day of the three-day election, the *American* implored every businessman to vote on his way home from work: "Is his dinner for a *single day* of more consequence to him than good water during a life time?" The editors needn't have fretted. A late-winter blizzard kept turnout low, but the proposition was approved in a landslide, 17,330 to 5,963.

A final reminder of the need for abundant water came the following winter. On a subzero night of blustery winds, hot coals in a Pearl Street warehouse ignited gas escaping from a broken line. A watchman immediately smelled the smoke, but the flames spread so fast that fifty square blocks were burning in fifteen minutes. People could see the glow from New Haven and Philadelphia. Thirteen acres of prime Manhattan real estate burned, including most of the old Dutch city south of Wall Street. Firemen, already exhausted by two blazes the night before, stood by in a helpless stupor. Their water was frozen in their hoses. But even on a summer night they couldn't have done much against this conflagration. Water for fighting fires in New York was so scarce the fire companies often fought each other with their fists to get control of the hydrants.

"GRAVE AND DIFFICULT MATTERS . . ."

The city hired the distinguished engineer David Bates Douglass to build an aqueduct for conveying Croton River water to the city. A graduate of Yale and West Point, he had served with honor in the army's elite Corps of Engineers during the War of 1812, then returned to West Point to teach. Most recently he had built the Morris & Essex Canal across New Jersey and run the survey for the Long Island Railroad. He was also an instructor at the University of the City of New York.

Douglass and several aides tramped up and down Westchester County all that summer, tracing and retracing possible routes. They saw a farm county of 36,000 widely scattered people, fully recovered from the battles of the revolution and prospering. On a map, the county looked like a funnel squeezed between the Hudson River and Long Island Sound, with its southern tip pointing at Manhattan. The Croton, the Saw Mill River, the Bronx River, and smaller streams cut steep valleys through farmland and forested hills on their

"The Great Fire of the City of New York," lithogaph, Hoffy, J. L. Bowen, and H. R. Robinson, 1835.

"The Great Fire of 1835," lithograph, H. Sewell, c. 1835.

"View of the Great Fire in New York, December 16th & 17th, 1835," colored aquatint, Nicolino Calyo, William J. Bennett, and L. P. Clover, c. 1835.

way to the Hudson. A few Dutch families whose manors had descended through many generations still dominated the region. Westchester was no wilderness in 1835, but it was certainly remote, a place that kept to itself, following old ways in quiet, pastoral comfort—the antithesis, in short, of its bruising urban neighbor. "If the Canadians should cross into our territory," one of the engineers cracked, "and plunder and destroy everything before their advancing to this country, the inhabitants here would not find it out soon enough to retreat." Indeed, the locals greeted the aqueduct builders from the city with little more hospitality than they would have offered a foreign invader.

Douglass had to choose the best place for a dam and reservoir; stake out the best route for the aqueduct itself (with politically delicate estimates of how much land must be purchased for the right of way); decide on good locations for storage reservoirs in Manhattan; and design the structures that would carry the aqueduct over rivers and through hills. He went at the work with diligence—perhaps too much of it. He was one of those men who always want a few more facts, a little more time, before deciding for sure. Thus his plans seemed never quite done, much to the consternation of the man who kept asking for them.

This was Stephen Allen, former mayor and state senator, now chairman of the powerful Water Commission. Allen was unlike the scholarly, stiff-necked Douglass in every way. While Douglass had entered the highest rank of his profession directly from the best schools, Allen had climbed to a position of influence from the low end of society. As a kid during the Revolutionary War he had been a sail-maker's apprentice on the South Street piers. He built on that ex-

New York's Great Fire began on December 16, 1835, seven months after municipal leaders had approved a city water system but long before the Croton Aqueduct was built.

perience to create the city's biggest sail-making business and a sizable fortune. In politics he was a Democrat; in his relations with others, "rude and unpolished . . . just, but not generous," honest, tough-minded, self-reliant, opinionated, and very hardworking. He did not like David Douglass at all, and the feeling soon was mutual. Allen would bawl out Douglass for unnecessary delays or "much lack of energy" or excessive spending. Douglass would complain, sometimes in public, about Allen's "uncivil [and] insulting language" or "sneers and rudeness." Allen retorted: "Because I am unable to clothe my observations in the feminine utterance of a learned sophist like Mr. D. am I to be charged with rudeness?" No one was surprised when Allen and his fellow commissioners fired Douglass in the fall of 1836.

In place of Douglass Allen wanted "a practicable engineer," someone who could make a big idea happen on schedule. As his feud with Douglass worsened, he scouted for such a man among the engineers who had built the Erie Canal, the technological achievement that loomed over all others of its day. The great canal, constructed by an army of laborers between 1817 and 1825, stretched from the Hudson River to the Great Lakes—a 363-mile trench through rugged, wild terrain, with 83 locks to climb hills and 18 aqueducts to cross rivers. The canal became an impromptu school of civil engineering, producing a generation of engineers who went on to build many of the nation's other canals and early railroads.

The star of the Erie Canal school was John Bloomfield Jervis. His father, a carpenter and farmer, had left Long Island in 1798 to settle at Fort Stanwix in central New York. Many New Englanders and New Yorkers saw a great future for the settlement outside the fort. It stood on the Mohawk River, where the river pierced the Adirondacks—the only water-level route through the Appalachian Mountains, and thus a sure trade route to the burgeoning West. The village took the hopeful name Rome, and in 1817 it became the point where construction of the Erie Canal began. At twenty-one, John Jervis watched the work with fascination. Soon, thanks to a word from his father to the engineers, he signed on as an axman, felling trees as the engineers did their surveys. He taught himself surveying and moved up quickly, taking charge of a seventeen-mile section of canal construction in central New York. He became chief assistant on the Delaware and Hudson Canal, then moved to railroads, building such early lines as the Mohawk and Hudson. In the

In 1832, former mayor Stephen Allen was named to the Water Commission. His appointment came as a surprise: he was politically unpopular and rough around the edges. But his reputation for honesty prompted his fellow water commissioners to name him chairman.

THE BIG DITCH

The West was America's future. From the revolution onward, few doubted it. But what role could New York, a city of ships, play in a drama that would materialize on a great land mass? The city's tendrils of trade reached across the Atlantic and around the world, but to the west, trade ran smack into the wilderness wall of the Appalachian Mountains. It cost more in 1810 to ship a ton of goods just thirty miles inland from New York than from the South Street piers to London.

Yet beyond the mountains, a treasure beckoned to New Yorkers. Beginning in the 1790s, a swelling stream of pioneers left New England and the Hudson valley for the rich lands seized from the Iroquois in western New York State and Ohio. These farmers sent their produce by wagon to New York for shipment along the Atlantic coast or overseas. But the mountain journey was bone-shakingly hard, and easier routes beckoned— by turnpike from Pittsburgh to Philadelphia; by the National Road to Baltimore; and especially by water down the Ohio and Mississippi rivers to the Gulf of Mexico. That burgeoning route threatened to rob New York of its trade supremacy and hand it to New Orleans, which was anyone's bet to become the nation's dominant city.

DeWitt Clinton saw the danger, and in the danger he saw opportunity—for himself as much as for his city. Arrogant, mercurial, and enormously ambitious, Clinton was the political boss and mayor of New York. He brutalized his enemies and bullied his friends; even his own secretary despised him. Yet Clinton had the wit to embrace a staggering scheme for an artificial waterway to the West and make it his own. The idea, which had originated among wealthy New York speculators in western land, encompassed a canal some 350 miles long—the world's longest by far—reaching across New York State from the Hudson River at Albany to the Niagara River at the young village of Buffalo. It would have to traverse swamps, dense forest, ravines, rivers, and rocky ridges. But if it could be built, it would allow freight-laden barges to skim along a water route at a fraction of the cost that land routes demanded. It would unlock the riches of the American interior and bring them streaming through the city of New York.

Even if they liked the idea, people who so much as glanced at the cost estimates just laughed. At $6 million, the price was equal to three quarters of the entire federal budget. Thomas Jefferson, certainly no enemy of westward expansion, reminded Clinton that Congress had rejected George Washington's plan for a 30-mile canal through Potomac River farmland because it would cost $200,000—"and you talk of making a canal *three hundred and fifty miles through the wilderness!* It is a splendid project, and may be executed a century hence. It is little short of madness to think of it at this day."

Clinton was mad enough to keep at it. Rebuffed by the federal government, he took his idea to the people of New York, talking endlessly about the canal's benefits to the city. The canal campaign became joined to Clinton's campaign for governor. Both were victorious, thanks to Clinton's ingenious plan to fund the canal's construction with bonds backed by a tax on salt, steamboat travel, and land within twenty-five miles of the canal route—a fiscal innovation that soon was copied across the country.

DeWitt Clinton, governor of New York, 1817–22 and 1825–28, and the driving force behind the Erie Canal.

When Gouverneur Morris, a canal backer, was chided about the physical obstacles the canal must surmount, the classically educated statesman coolly replied: *"Labor improbus omnia vincit"* ("Labor overcomes all obstacles"). That was easy for a soft-handed aristocrat like Morris to say. In the case of the canal, it turned out to be perfectly true. The work was let to dozens of individual contractors, each of whom built his own small section under the supervision of canal engineers paid by the state. The contractors, in turn, hired their own laborers. Many were farmers and townsmen who lived close to the route. Others were Irish and Welsh immigrants hired off the wharves of New York. Thousands were employed at a time. For a top wage of fifty cents a day, they built the greatest public works project since the pyramids using hand tools and muscles— their animals' and their own.

The middle section—the ninety-four miles between the Mohawk and Seneca rivers—was built first. Construction began with a ceremonial shovelful of dirt and the boom of cannons at Rome on July 4, 1817.

The work went forward under men who had little or no formal training as engineers. They taught themselves as they went along. They found, for instance, that the traditional European method of canal digging, using spades and wheelbarrows, was easily bettered by teams of three men with horses or oxen pulling a plow and scraper. The need to clear the route of thousands of trees led to a clever device whereby one man could fell a tree without an ax or saw: He twisted a large screw into the ground a hundred feet from the tree, tied a cable from the screw to a high point on the trunk, then cranked the cable until the tree crashed down. Another machine improvised on the spot allowed seven men and a team of oxen to pull thirty or forty stumps from the ground in a day. To prevent leaks, a young engineer named Canvass White concocted a solution of powdered limestone, water, and sand that would harden under water; some 400,000 bushels of the stuff were eventually used on the canal. The engineers' techniques revolutionized canal building. After four years, one of them wrote: "For accuracy, despatch and science, we can now present a corps of engineers equal to any in the world. . . . The canal line is now one of the most excellent schools that could be devised, to accomplish men for this pursuit."

Simple digging was the main work on the central section. To the west, great structures had to be built. To get the canal across the low valley at Irondequoit Creek, the engineers constructed an embankment of stone and earth 70 feet high. The canal ran across the embankment; the waters of the creek ran through it via a culvert 245 feet long, supported by some 900 piles pounded into the valley's quicksand. In the swift-running waters of the Genesee River, the canal builders erected 9 stone arches to make a river-spanning aqueduct more than 800 feet long—one of the greatest of the canal's 18 aqueducts. It became the central edifice in the boom town of Rochester.

The most daunting obstacle was a mountainous ridge that rose near Buffalo, the western

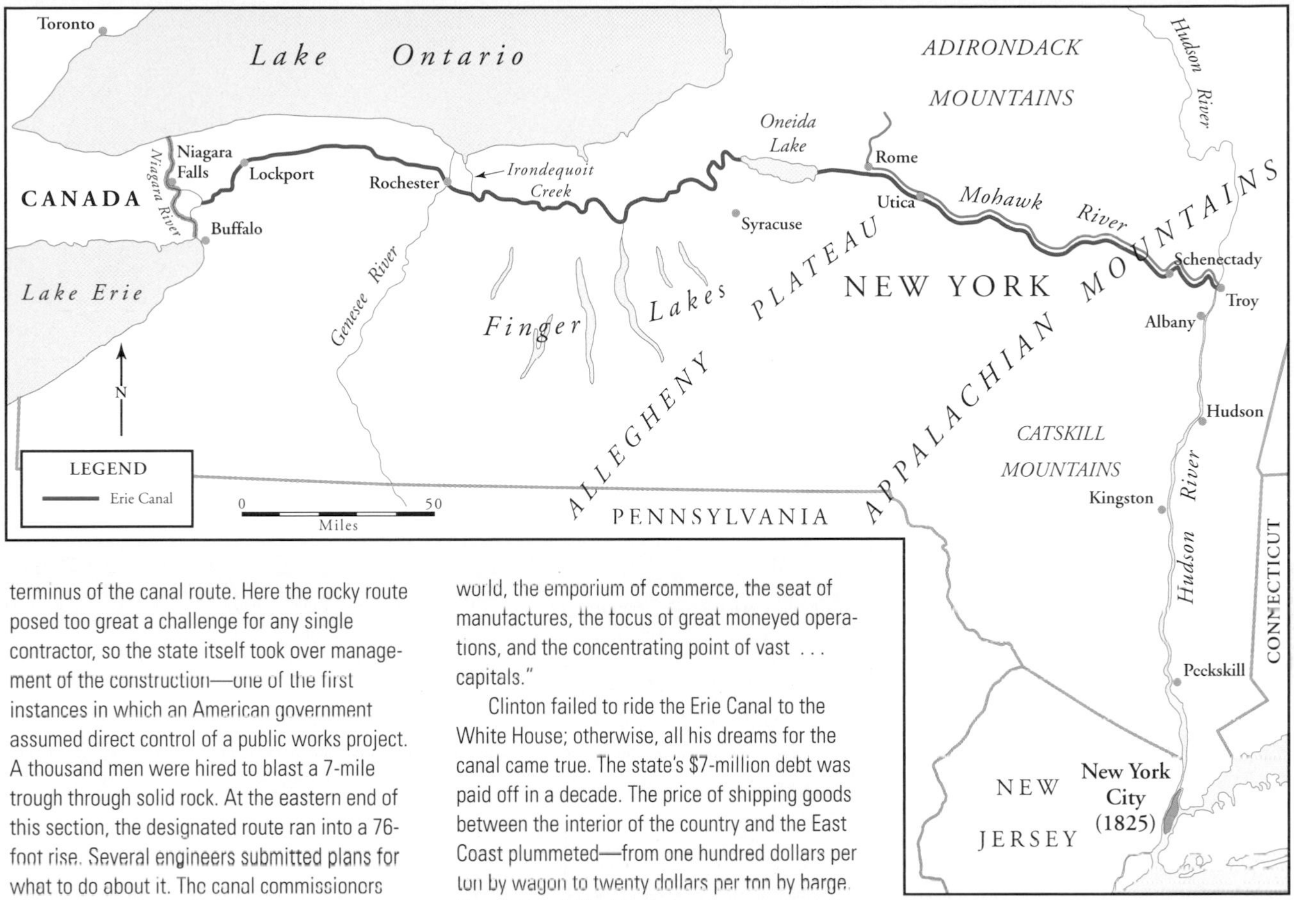

The canal's route, from the Hudson River west to Buffalo and the Great Lakes.

terminus of the canal route. Here the rocky route posed too great a challenge for any single contractor, so the state itself took over management of the construction—one of the first instances in which an American government assumed direct control of a public works project. A thousand men were hired to blast a 7-mile trough through solid rock. At the eastern end of this section, the designated route ran into a 76-foot rise. Several engineers submitted plans for what to do about it. The canal commissioners chose the design of a stern, exacting engineer, Nathan Roberts, who envisioned two great flights of five navigational stair steps, one flight for the westward route, one for the eastward. Using wood, iron, and fifty thousand square feet of facing stone, the workers took four years to build these locks at the new town of Lockport, the last and most spectacular structure of the entire canal.

Upon its completion in the autumn of 1825, the canal was 363 miles long, 4 feet deep, and 40 feet wide. It accomplished a combined ascent and descent of 675 feet via 83 locks. It was declared complete in a great celebration that climaxed with Governor Clinton's ceremonial pouring of two kegs of fresh water from Lake Ontario into the saltwater of New York harbor. "As an organ of communication between the Hudson, the Mississippi, and the St. Lawrence, the great lakes of the north and west, and their tributary rivers," Clinton declared, the Erie Canal "will create the greatest inland trade ever witnessed." That trade would flow through New York City, he said, making it "the granary of the world, the emporium of commerce, the seat of manufactures, the focus of great moneyed operations, and the concentrating point of vast . . . capitals."

Clinton failed to ride the Erie Canal to the White House; otherwise, all his dreams for the canal came true. The state's $7-million debt was paid off in a decade. The price of shipping goods between the interior of the country and the East Coast plummeted—from one hundred dollars per ton by wagon to twenty dollars per ton by barge. New York solidified its trade supremacy over New Orleans and became the nation's leading mercantile center for good. Meanwhile, hundreds of thousands of emigrants streamed westward through the canal, paying a penny and a half per mile for the easy eight-day trip. These travelers made great cities out of the frontier outposts of Buffalo, Cleveland, Detroit, and Chicago. At its peak the Erie employed a workforce of 50,000 and spawned a subculture celebrated in story and song. It inspired a 3,300-mile network of canals that soon girdled the East from Indiana to Boston—the nation's first mass transit system—though none approached the Erie in importance. The canal knitted the Midwest to the Northeast, an alliance that would win the Civil War. By then the canals were surpassed by a new and mightier technology, the railroads. But the Erie Canal had proven beyond any argument that great public works were central to the nation's future.

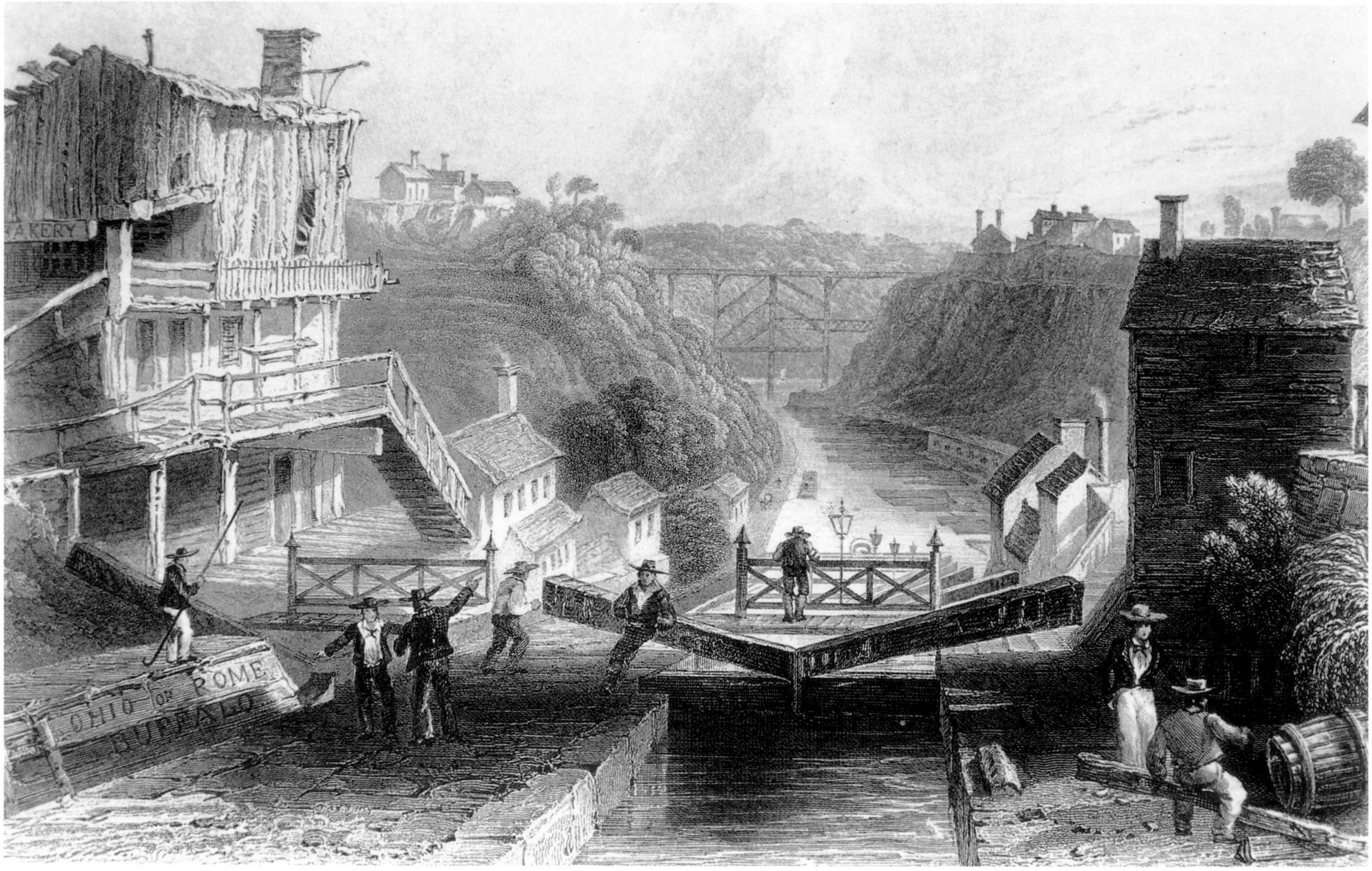

"Lockport, Erie Canal," engraving, W. Tombleson after W. H. Bartlett, c. 1830. **John Jervis began to build his reputation with the Erie Canal.**

fall of 1836, recently widowed at forty-one, he was making improvements to the Erie Canal when Stephen Allen came calling.

Jervis breathed the air of New England Congregationalism, with its Puritan devotion to good works and disdain of idleness. He was a master of detail, taking pride in the neatness of his accounts and records, though not so much that it slowed him down. His biographer reports that "he savored every moment as one that could and would be used for some form of industry." He was utterly absorbed in his work. "A true engineer. . . ," he wrote, "considers his duties as a trust and directs his whole energies to discharge of the trust. . . . He is so immersed in his profession that he has no occasion to seek other sources of amusement, and is therefore always at his post. He has no ambition to be rich, and therefore eschews all commissions that blind the eyes and impair fidelity to his trust." He was also a Democrat, which didn't hurt his standing with Stephen Allen—though for the rest of his career Jervis had to beat down the rumor that he won the Croton job through political intrigue. Certainly Allen liked him, especially in contrast to the persnickety David Douglass.

"The enterprise of the Croton Aqueduct was an improvement

for which there was no specific experience in this country or hardly any in modern times," Jervis wrote later. "It was hydraulic, and in this respect resembled canals; but it had peculiarities that had no parallel in canals." As on the Erie Canal, he would have to improvise. For guidance, he had only a couple of books on hydraulic engineering, both published in Britain. In 1836 there was not yet a single American treatise on civil engineering. Jervis knew, too, that the standards were very high. On a canal or a railroad, an engineering flaw meant only a suspension of traffic. But a flaw in the aqueduct would threaten the daily safety and health of a great city, and "any material failure would be disastrous in the extreme." With these things in mind, he wrote later, "I commenced the work with confidence that I could bring it to a successful result; at the same time I felt the work involved very grave and difficult matters, and entered on the enterprise deeply impressed with the responsibility of the undertaking."

John Jervis at mid-career.

The task of an aqueduct builder is to make rough places smooth and crooked places straight, all to accommodate a pipe. To get water from one place to another, the pipe must run down a smooth, gradual slope that a kid with a wagon could navigate without once having to push himself uphill. The topography of Westchester County would make this difficult indeed. It was "the meanest country that I ever met with," a surveyor said, and the Water Commission agreed. "We are met at every step with deep ravines," the commissioners said, "which must be passed, either by embankment or bridge; or elevated hills, which must be pierced by a tunnel of more or less extent."

Jervis discovered that David Douglass had done little more than lay out a route from the Croton to the edge of Manhattan. ("From the long time he had been engaged on the work I did expect to find more progress in preparation," Jervis said later.) That much had taken Douglass eighteen months. All the planning that remained took Jervis eighteen *weeks*—designs and specifications for a dam and reservoir on the Croton; for sections of aqueduct that would run for many miles just above or below the surface of the ground, depending on the terrain; for arches and embankments to carry the pipe over rivers and streams; for tunnels to take it through hills; for a bridge from Westchester to Manhattan over the Harlem River; and for a system for distributing the water in the city, including giant standing reservoirs. Jervis swiftly divided the job into ninety-six sec-

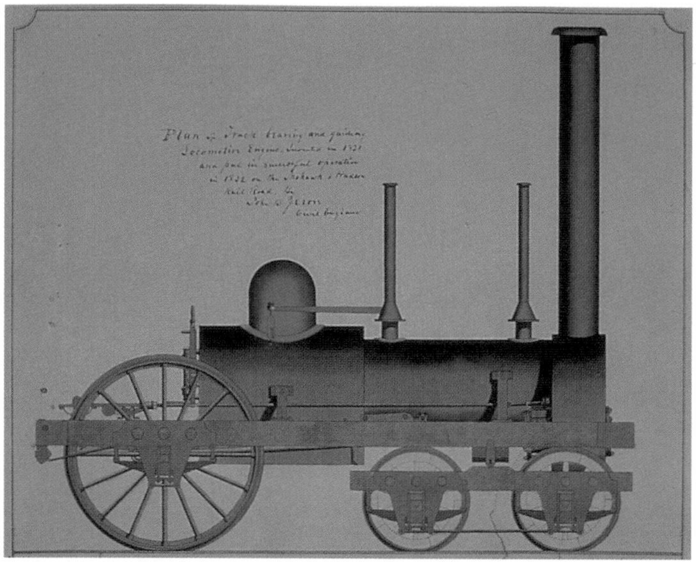

Jervis directed the construction of the Mohawk and Hudson Railroad in 1830, the first railroad line in New York and the second in the nation. Here he has painted an early locomotive design for that line.

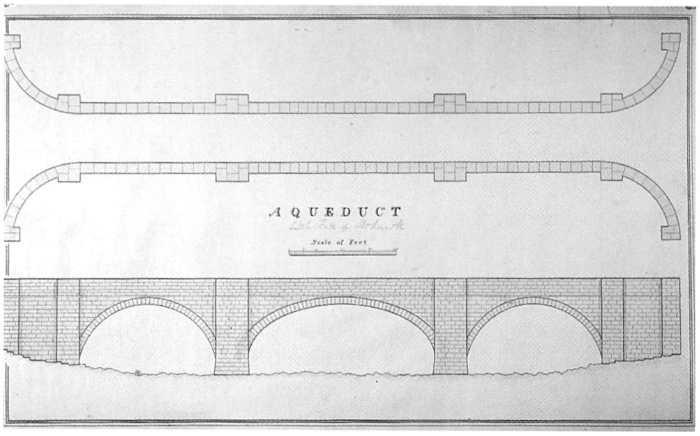

"Canal Aqueduct over the Mohawk River at Little Falls," ink and watercolor, John Jervis.

tions, with contracts awarded to the low bidder in each section. With a deadline of October 1839, just three years away, he wanted no time wasted. He retained Davis's aides but told them to get their men moving at the crack of dawn. "The days are short," the new chief warned, "and to make much progress in the field work it is indispensable to have an improvement of their early hours."

As the surveyors marched through Westchester, some two hundred landsmen whose property would be flooded by the dam or crossed by the aqueduct started a running defensive battle. When Jervis's men paid their legally required calls to negotiate land transfers, the Westchesterites were unaccountably away. They refused offers for their land, forcing the Crotonites to go to court for lengthy condemnation proceedings. They swiped surveyors' stakes. They sued. They cursed the aqueduct men and roughed them up. Some of them speedily divided their farm acreage into phony village lots and jacked up their asking prices accordingly. They didn't stop the aqueduct, but they were handsomely compensated for their trouble. Stephen Allen believed the city paid $257,000 for land that was actually worth $65,000.

At the north end of the project, the work began with the fifty-foot-high Croton Dam, six miles up the Croton from the Hudson. The dam backed the river up into a five-mile-long reservoir with a capacity of 500 million gallons. From this great holding tank, water would enter a channel in the southern abutment of the dam and begin the gentle descent to the city, first along the south bank of the

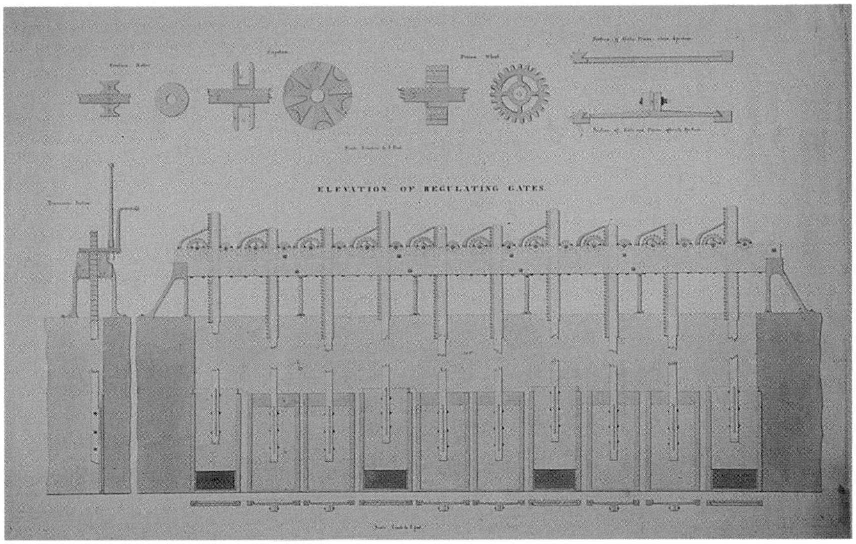

"Regulating Gates to Control Flow into the Croton Aqueduct," ink and watercolor, John Jervis.

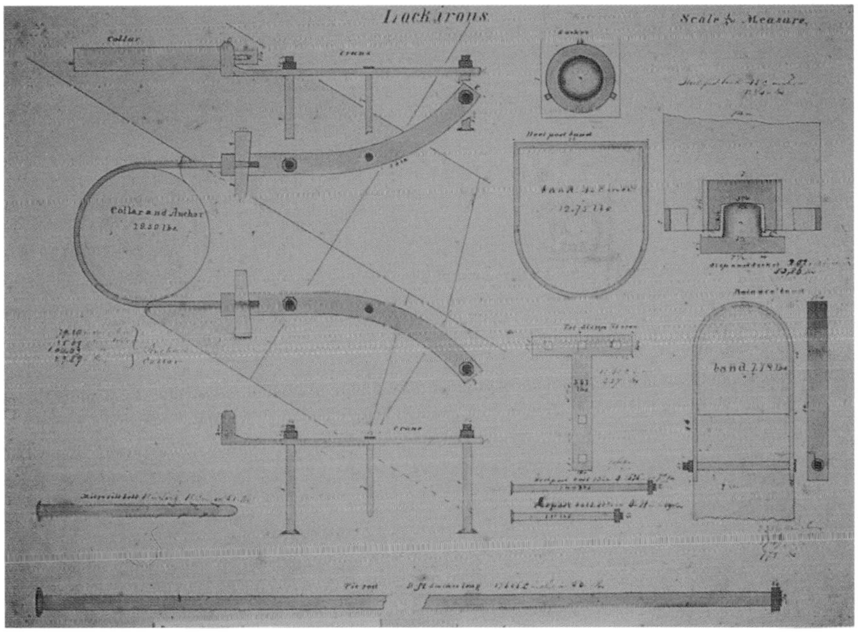

"Lock Irons," ink and watercolor, John Jervis.

Croton, then along the east bank of the Hudson through Sing Sing, Tarrytown, Dobbs Ferry, Hastings, and Yonkers, then inland across the Saw Mill River in a curve through the Bronx to the Harlem River.

To take the flow, Jervis designed a pipe 33 miles long. Its cross section was horseshoe-shaped, 8½ feet high and 7½ wide. The floor of the pipe, resting on a stone foundation, was an inverted arch made of 8-inch-thick brick. The walls were made of crushed stone mixed with hydraulic cement. Most of the aqueduct was built on the "cut and cover" plan. That meant the pipe was laid in a trench, then covered with earth, leaving only a long berm, or mound, to show those at ground level where the aqueduct was, save for small stone towers that rose here and there to provide ventilation for the buried pipe.

Immense physical labor was required, most of it done by some three thousand unskilled workers. Most were Irish Catholic strangers in a Protestant land, and they were made none too welcome in Westchester. Drink was a major issue. Project rules forbade the sale of liquor to workers, but some Westchesterites, spotting a captive and thirsty market, evaded them. "The love of lucre," the water commissioners scolded, "has induced certain individuals, regardless of the injury inflicted on others, to open places of resort for the laborers, where this *enemy of man* may be obtained, in any quantity, for money." Some local farmers thought the laborers "a civil people," but most thought they were objectionable from every angle—moral, behavioral, linguistic, and hygienic. An outraged Joshua Purdy submitted a bill for damages in the amount of $3,012. He said the charge was only for quantifiable losses caused by workers, not "the inconvenience, trouble and anxiety of having between three & four hundred Irishmen upon my own farm. . . . I can assure you it is no pleasant thing to have . . . huts or shantees . . . stuck up within a few rods of my dwelling and peopled with the lowest and most filthy of mankind."

REPAIRING THE BROKEN PIPES OF THE CROTON WATER MAIN IN FIFTH AVENUE, AT SIXTY-FIFTH STREET.—SEE PAGE 74.

"Cut and cover" tunneling on the aqueduct line.

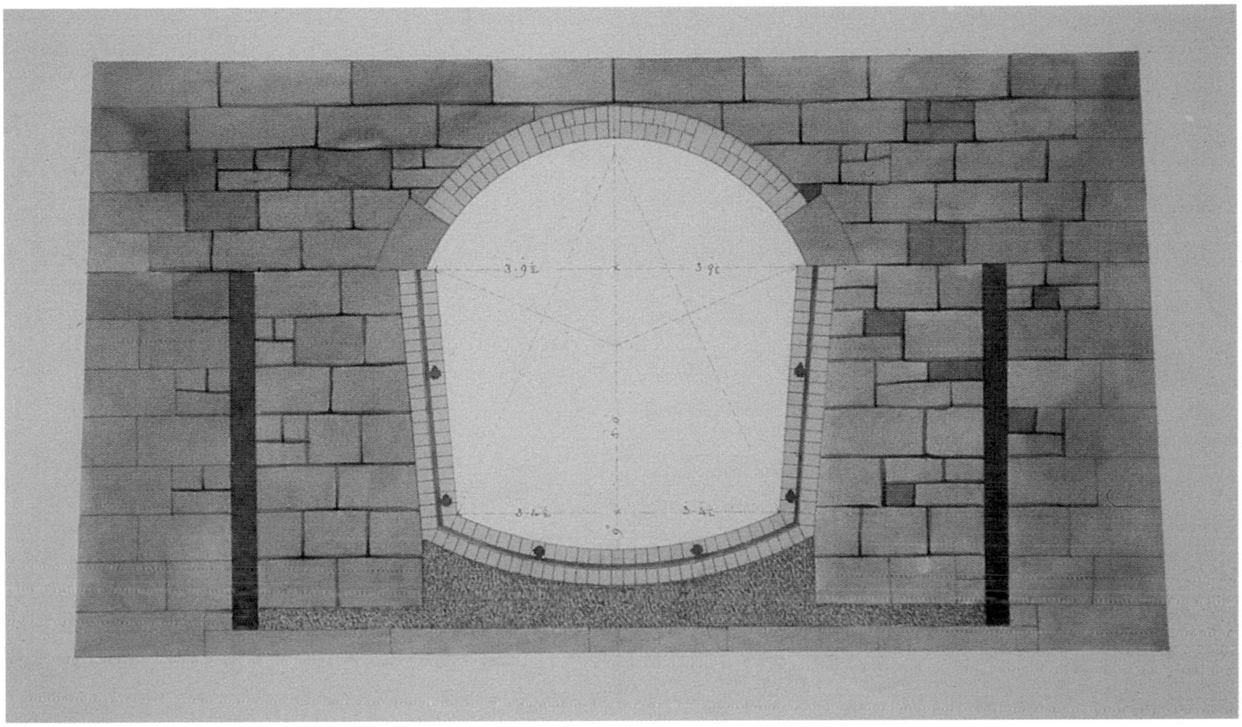

Jervis's watercolor shows his plans for the Sing Sing arch. To prevent leakage from the waterway and freezing in the supporting structure, Jervis followed the example of Scottish engineer Thomas Telford, who lined a Welsh aqueduct with iron.

The most famous resident of the aqueduct line was also the most patronizing toward the Irish. In the village he immortalized in *The Legend of Sleepy Hollow,* Washington Irving wrote of Irish laborers "spooked" by familiar headless phantoms: "In a word, the whole wood has become such a scene of *spuking* and *diablerie,* that the paddys will not any longer venture out of their shantys at night, and a Whisky-shop, in a neighboring village, where they used to hold their evening gatherings, has been obliged to shut up, for want of custom." Without drink, Irving taunted, the workmen might "entirely abandon the Goblin regions of Sleepy Hollow, and the completion of the Croton Waterworks [may] be seriously retarded."

While Irving scratched his droll stories, the Irishmen built a prodigious public structure. The route of the aqueduct twisted and turned to follow the proper gradient. Sometimes it ran into a ridge that couldn't be gotten around. Here Jervis set the men to digging tunnels. There were sixteen in all on the aqueduct, including one of 640 feet through Asylum Hill in Manhattan. The tunnels were immense jobs in themselves. At one place it took laborers six months, working in round-the-clock shifts, to dig a 375-foot shaft. When the route crossed the valley of a river or creek, Jervis had to decide whether to build a bridge over the river, with the aqueduct

"Aqueduct Bridge at Sing Sing," engraving from *Illustrations of the Croton Aqueduct,* Fayette Tower, 1843.

running atop it to the other side, or to pile up a big embankment in the valley, with culverts through the embankment to carry the river and the aqueduct running along the embankment's ridge. This invariably demanded a difficult judgment based on complex assessments of cost.

In aesthetic terms, the residents were always better off with bridges, for Jervis's designs were elegant. To cross the 500-foot-wide valley at Sing Sing (now Ossining), he built a single elliptical arch bridge in a "plain and substantial style of architecture" that the *Westchester Herald* called "the most astonishing specimen of art and ability." Few people of the 1830s had ever seen a structure as massive as the Sing Sing Arch, or contemplated a work of man that stretched far beyond both horizons. It dawned on many that the aqueduct was more than a purely utilitarian contrivance for getting water to Manhattan. A work of enduring beauty, even a symbol of man's capacity to master his environment, was emerging in the Westchester countryside. "It is surprising," said Alexander Wells, an editor in northern Westchester, "to observe how beauty and solidity are blended in the construction of this stupendous work." In a letter to his mother in upstate New York, Fayette Tower, a young engineer, said: "I think it will be visited by foreigners not only as a model but as an illustration of what the ingenuity of man led on by the pure light of science can accomplish, and they will admire the gigantic undertaking and the boldness of conception."

Greater works of masonry had arisen in history, but mostly under duress, like the Egyptian pyramids, or from military necessity, like the Great Wall of China. This seemed different. The Water Commission took note that "no population . . . ever before voluntarily decreed that they would execute such a work. No population but one of freemen would have conceived the idea." Its purpose was not "protection from external foes," the commissioners added;

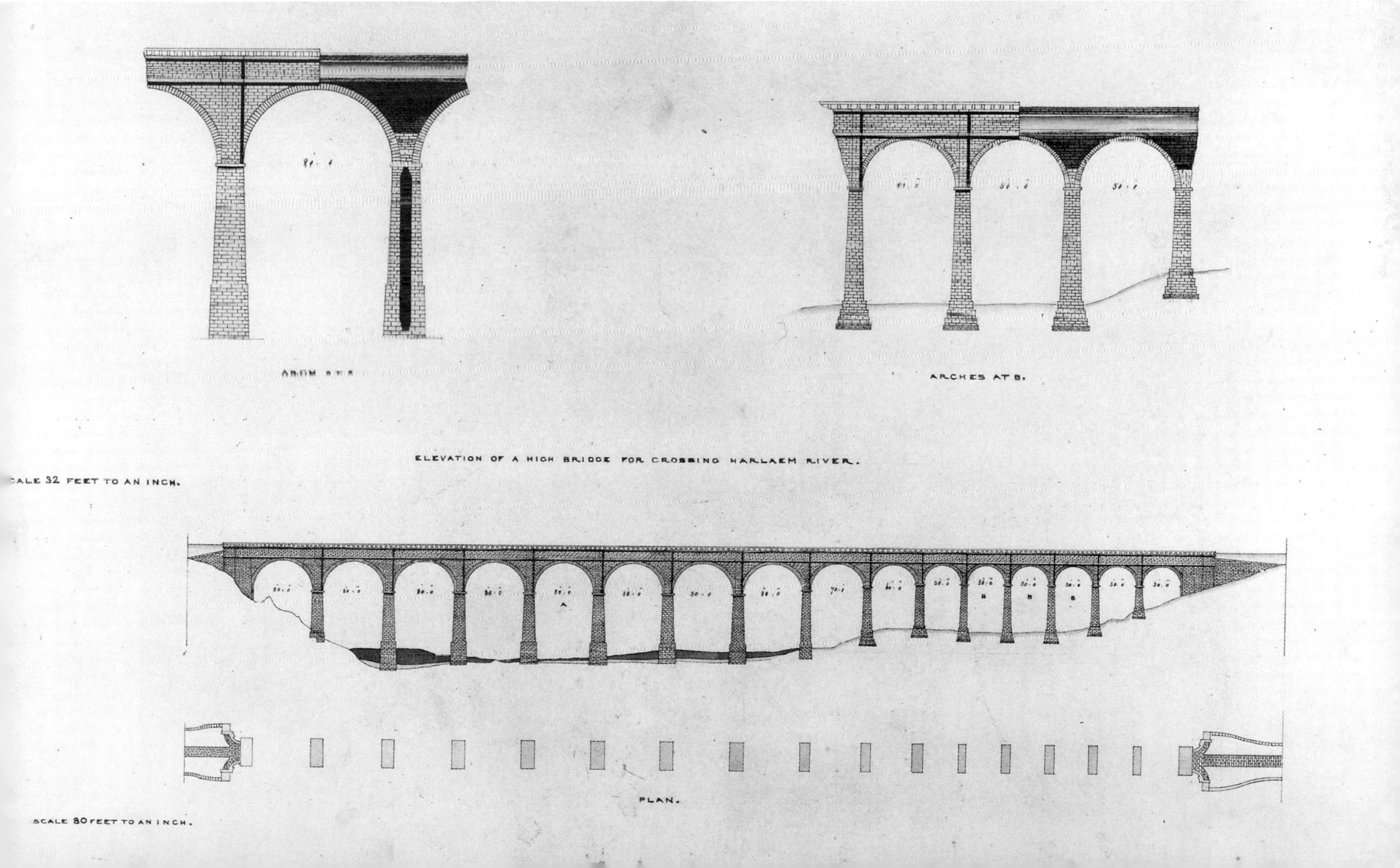

"High Bridge for Crossing Harlem River," ink and watercolor, John Jervis.
The bridge's arches were to be at least 80 feet wide, providing at least 100 feet of clearance at high tide.

"View of the High Bridge," engraving, 1861.

rather, "it looks to making our whole population happier, more temperate and more healthful, and does contemplate that the countless millions hereafter to enjoy the benefits of this water, will have clear heads, correct eyes, strong arms, and instead of walls, present breasts so strong and hearts so brave, that in a just cause our city may defy all foreign foes."

Ironically, the Westchester men who had opposed the aqueduct at every opportunity became the ones responsible for its most spectacular structure. The question was how to get the pipe across the Harlem River to the city. For the aqueduct to span the river at the level of its gradient, Jervis would have to build the tallest bridge in the United States to date. Jervis thought such a bridge would cost nearly a million dollars, whereas he could build a small, low bridge or embankment—with pipe siphons to suck the water into Manhattan—for half that cost. But here the Westchesterites made their final stand. The shallow Harlem was hardly a thriving commercial route, but they insisted a low bridge would impede future boat traffic. Manhattanites suspected that "gentleman speculators" from Westchester were behind the push for a high bridge; they had bought up land on both sides of the Harlem in hopes of a bridge big enough to accommodate people as well as water, which would send the value of their property soaring. But whatever their motives, the Westchester forces got their way in Albany, where the legislature forbade a low bridge. When the fight was over, John Jervis was not so disappointed. "As you know," he told a friend in the profession, "engineers are prone to gratify a taste for the execution of prominent works." The bridge took several years to complete; a low embankment was built to carry the water across the river while the bridge was under construction. When it was done, the High Bridge, 1,450 feet long with 15 100-foot arches, became a landmark that still stands.

"High Bridge and Reservoir," duotone lithograph from the *Valentine Manual,* 1859. **This impressionistic illustration ignores geography in its depiction of the High Bridge bringing water to the York Hill Reservoir in central Manhattan.**

The aqueduct entered the island of Manhattan at 173rd Street, not far from the mansion of Madam Elizabeth Jumel, long the mistress and briefly the wife of Aaron Burr. The street number suggests the urban landscape of later years, but in 1840 the northern portion of Manhattan remained a country place. Thus, even in New York, the aqueduct passed through broad farm fields, past grazing livestock, as it ran south for a mile to the village of Manhattanville, through Manhattan Hill in a 1,215-foot tunnel, then across a strikingly beautiful land bridge in the Clendening Valley, nowadays part of the Upper East Side.

At York Hill, the future site of Central Park, Jervis built a giant receiving reservoir to take in the Croton treasure. It was 1,826 feet long and 836 feet wide, with massive, sloping walls to hold 150 million gallons. Three pipes ran from York Hill east to Fifth Avenue, then south for some fifty blocks to a second, smaller reservoir at Fifth and Forty-second Street. From these, the water would be distributed to residents throughout the city. Nearly 50 feet tall and 420 feet square, this reservoir was built as a triumphant symbol of the new system, with Egyptian-style ornamentation. Passersby could climb a stone staircase to the top and walk along the edge, looking down into water that had traveled forty miles and beyond it to the panorama of the city. The reservoir's "compact and massive walls, seemingly strong and enduring as the everlasting hills, with their quaint Egyptian cornices, give to the vast edifice a very sphinx-like aspect," said a writer in *Harper's* magazine a few years later. "No citizen passes it, pass often as he may, without a vivid consciousness of its presence, and no stranger fails to ask curiously of its character and purpose." From here, 167 miles of iron pipe snaked out through the city, feeding offices, factories, hydrants, and homes. (By Jervis's day the site of the receiving reservoir was named Murray Hill, for John Murray, the Quaker merchant who had first owned the property. New Yorkers, ever impatient to replace old structures with new, would build the New York Public Library on the site of the receiving reservoir at the turn of the twentieth century.)

Croton water was first sent to New York with the nineteenth century's characteristic devotion to ceremony. The pageantry began at 5 A.M. on June 22, 1843, when John Jervis ordered water released from the Croton River into the aqueduct. A flow about eighteen inches deep went on its way. Four men set a long wooden skiff called the *Croton Maid* in the pipe and hopped aboard. In several legs, they rode the flow down to the Harlem River crossing, through the underground pipes and over the bridges, with Jervis himself and three members of the Water Commission taking the final leg. The engineers calculated that water was taking about twenty-two hours to arrive from the Croton. They paused, allowing the flow to run off through a spillway into the Harlem. On June 27 they closed the spillway and sent the water down the aqueduct to the Yorkville Reservoir, where Governor William Seward and a crowd of 20,000 heard thirty-eight guns salute the first freshet of water spilling across the floor of the reservoir. The aged John Jacob Astor, the richest man in America, watched with the others. "I think you ought to make *money* here," he told one of the contractors.

On July 4, the flow was sent on to the receiving reservoir at Murray Hill. Forty-five cannons boomed as the first water spilled into the tanks. It was so early in the morning that not many were there to see that it was a little muddy; the first currents had washed the dirt and sand out of the new pipes. The engineer Fayette Tower recorded the scene: "At an hour when the morning guns had roused but few from their dreamy slumbers, and ere yet the rays of the sun had gilded the city's domes, I stood on the topmost wall of the reservoir and saw the first rush of the water as . . . [it] entered the bottom and wandered about, as if each particle had consciousness." Some 25,000 visited the reservoir by evening. All were offered a cup of the Croton.

There was an uncanny moment late in the day. Mayor Robert Morris, worried about the danger of fire from Independence Day fireworks, asked that water be sent downtown to the old Thirteenth Street Reservoir and to the even older tank of the much maligned Manhattan Company. The water commissioners agreed. When the stream reached the Manhattan Company building, it quickly filled the three-story tank and overflowed, dousing the premises and shooting out the windows onto neighboring houses. "This was no

"Distributing Reservoir," John Jervis. **Jervis's plan for the Murray Hill distributing reservoir had sloping walls averaging forty-five feet high. He designed the walls in Egyptian style, with raised pilasters at the corners and a templelike entry. A cornice along the top of the reservoir walls would create a public promenade with an iron railing.**

In 1900, the Murray Hill Reservoir was demolished to make room for the New York Public Library and Bryant Park. The demolition process revealed the elaborate tunnel-work of the reservoir.

pleasant affair," said the reporter from the *Commercial Advertiser,* "and looked like satisfying an old grudge."

The final cost of the water system was $13 million, $8 million more than originally promised—not the last time that a major public works project, encountering unforeseen obstacles and political difficulties, would far exceed its budget. Yet "not a voice has been raised against it," wrote former mayor Philip Hone, "and all parties hail the advent of the 'pure and wholesome water,' after its journey on the earth, and under the earth, and across the watercourses of miles, as a proud event for our city, and one which enables the Knickerbockers to hold their heads high among the nations of the earth."

Two days before the final celebration that fall, Hone wrote: "Nothing is talked of or thought of in New York but Croton water; fountains, aqueducts, hydrants, and hose attract our attention and impede our progress through the streets. . . . Water! water! Is the universal note which is sounded through every part of the city, and

"Croton Water Celebration 1842"
Pomp and pageantry greet the arrival of clear Croton water on July 4, 1842.

infuses joy and exultation into the masses." On October 14, a quarter million people crowded Broadway in the city's greatest parade to date, marching from the Battery to Union Square, where a new fountain spouted Croton water in a dance of shimmering shapes.

For homeowners, the cost of installing their own pipes was high. So, although the price of water itself was only ten dollars per two-story household per year, household usage remained low for a time. After three years, only one in three residents had signed up for service, though many more took clean water from the free public hydrants. When John Jervis completed the High Bridge in 1848, water pressure improved. That brought more customers. When another cholera epidemic struck in 1849, people finally abandoned the old wells for good. By the 1880s, New Yorkers were using a hundred gallons of clean water per capita per day, the highest rate in the world.

Myndert Van Schaick, Stephen Allen, and John Jervis, all far-sighted men, believed they were building a system that would supply New York's need for clean water into the distant future. Yet even the most optimistic New Yorkers could not foresee the effects on the city of European famine and political upheaval; or of great technological systems such as the Erie Canal, which sealed the city's role as the nation's commercial capital; or of the aqueduct itself, which made the city cleaner, healthier, and more hospitable. All these things brought new people to New York in torrents. By 1860 the city was approaching a million residents, and once again there was not enough water.

So the race between supply and demand continued. By 1862, a new, billion-gallon reservoir had to be built in Central Park. From 1885 to 1893, an entire new aqueduct was laid from Manhattan to other parts of the Croton watershed, and another city reservoir was built at Jerome Park. Then, in 1898, Manhattan joined hands with Brooklyn, Queens, the Bronx, and Staten Island to form Greater New York, an urban giant of 3.5 million using 370 million gallons of water per day. Faced with this enormous demand, which was certain to grow, city and state leaders once again reached out a long arm for water, this time to the southern slope of the Catskill Mountains, 100 miles north of the city. There the great Ashokan Reservoir was built, 12 miles long, with a capacity of 130 billion gallons. Its water flowed to New York through the 92-mile-long Catskill aqueduct, another prodigy of civil engineering, which was finished in 1917 just as a water shortage in the city approached the level of an emergency.

Croton water was so popular that tradesmen incorporated it into advertisements.

As the twentieth century went on, the water system reached farther still into the Catskill range, beyond the Hudson watershed to the tributaries of the Delaware River, requiring decades-long negotiations with the states of Pennsylvania and New Jersey. More reservoirs and yet another aqueduct were built. To get the water into New York, two immense tunnels—forty miles in all—were dug through solid rock hundreds of feet below the Hudson and the city. A third city tunnel has been in the works, off and on, for a quarter century, often spurring debate as heated as the struggles that attended the birth of the first Croton aqueduct. Nearly all of New York's water still comes to the city by the simple force of gravity, without the help of pumps, thanks to the natural slope that John Jervis used to great advantage in the 1830s. Experts recognize the system as one of the

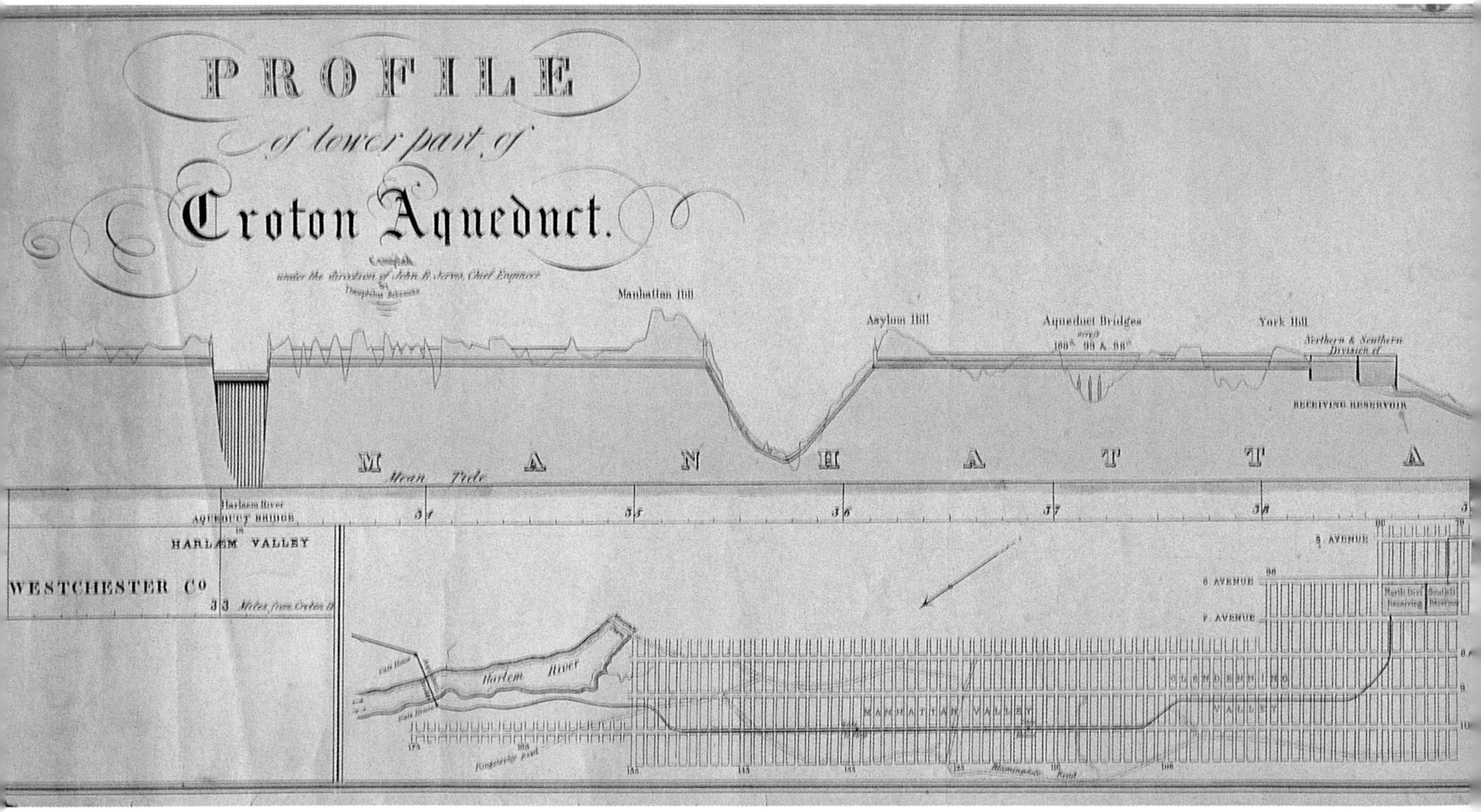

most efficient in the world—indeed, as one of the supreme triumphs of civil engineering.

The final acts in the saga of the old aqueduct came well into the twentieth century. The Manhattan Company's waterworks were torn down in the early 1900s, but the heirs of Aaron Burr's bright idea still worried about their state charter. So an employee pumped a little water at the site every day until 1923, just to keep the bank safe. The great pipe was closed in 1955. It lies empty under the ground near the Croton and Hudson rivers. People of the twenty-first century now jog on trails along the berm, dimly aware if at all of the story beneath their feet, or of the immense labors required to make a city.

A profile of the aqueduct line as it crossed Manhattan.

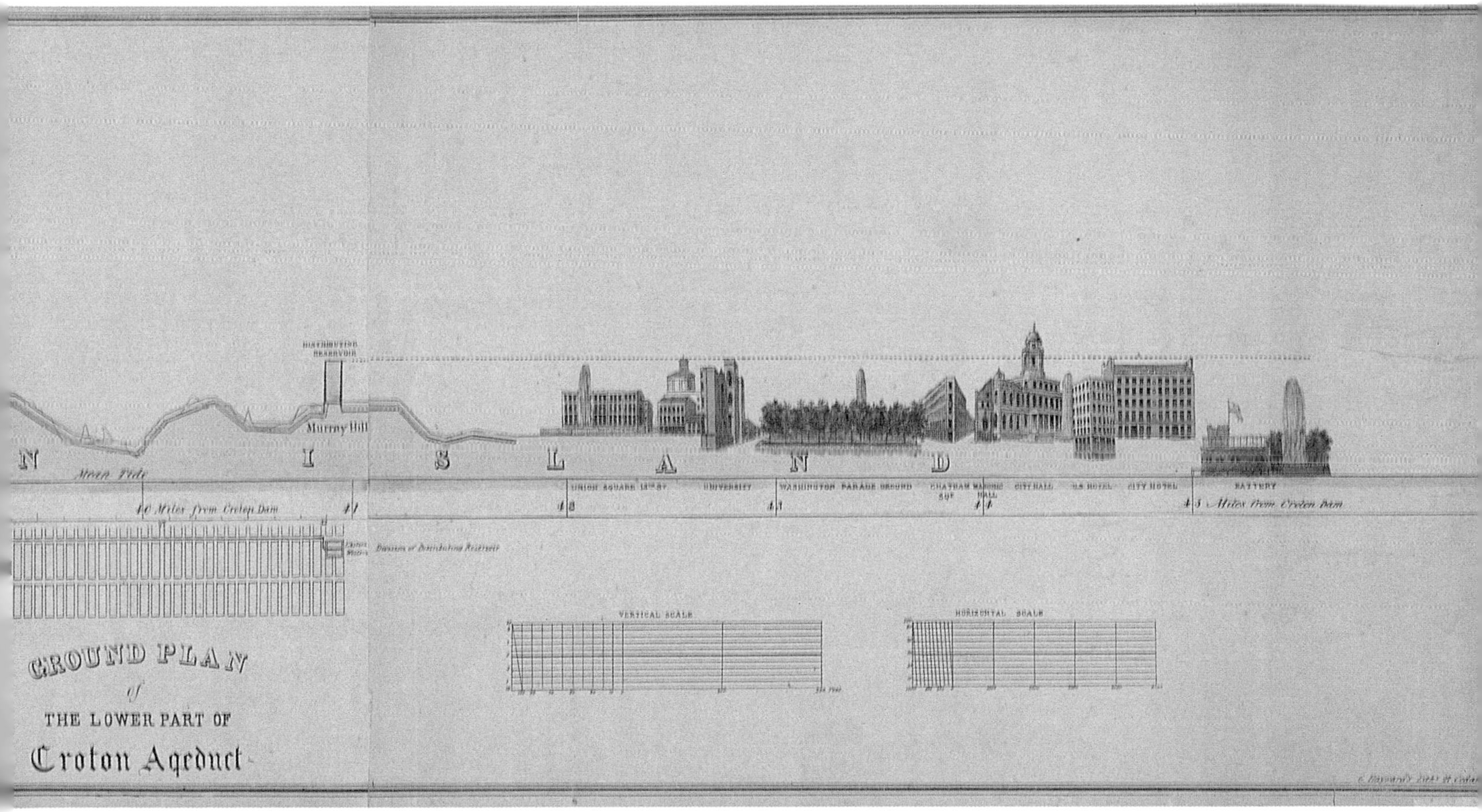

CHAPTER SIX

AMMANN'S BRIDGE

Keen instruments, strung to a
vast precision
Bind town to town and dream
to ticking dream.
– Hart Crane, *The Bridge*

Self-portrait, Othmar H. Ammann, 1904. **Ammann, age twenty-five in this photograph, had just arrived in the United States from Switzerland.**

When Othmar Ammann was twenty-five years old, he stepped on the precise spot where his destiny would be sealed many years later. It was a piece of ground on the New Jersey Palisades, the cliffs that overlook the broad Hudson River and the north end of Manhattan beyond.

Ammann—a slim, cultivated young man, "taciturn but passionate," a writer would later say—had been in the United States only a few weeks. Like many Europeans arriving in New York, he spent his first days ogling the hurrying crowds and "daring buildings." "So much has passed my eyes," he wrote his parents in Switzerland. "Everything is in a busy excitement and everyone is working for a goal without caring about the rest of mankind." He was not a tourist. He was an engineer, and a brand-new one, brimming with the hope of youth and enchanted by the sovereign structure of his chosen profession—the long-span suspension bridge. So, after he had examined the Roeblings' Brooklyn Bridge, he went up the Hudson River to the Palisades. There he could measure the landscape with his eyes and visualize the enormous project of the great Gustav Lindenthal, the preeminent bridge engineer of the day, who planned to throw a bridge twice as great as the Roeblings' across the Hudson. Ammann wrote later: "For the first time I could envision the bold undertaking, the spanning of the broad waterway with a single leap of 3,000 feet from shore to shore, nearly twice the longest span in existence. . . . For a young engineer it was a thrill to contemplate its possibility."

In time, Ammann's own mind would sketch a vision of another Hudson River bridge. The image would be just as vivid as the one in Gustav Lindenthal's mind, and Ammann's desire to build it just as great. But only one engineer could build the first bridge across the Hudson.

In hindsight, it would seem fitting that they shared this vision. Though they were born a generation apart, with utterly different temperaments, they followed strikingly similar paths to the Hudson River. Both learned German as their native language—Lindenthal in the Austrian Empire, Ammann in Switzerland. In education they differed. Lindenthal taught himself engineering as a carpenter and stonemason, while Ammann learned from the best professors. But they both soaked up the engineering traditions of central Europe, where the spread of railroads had nourished a venerable tradition of bridge building that reached back to the Renaissance. Both men hur-

"New York and Brooklyn," Currier & Ives, 1876. **After its completion in 1883, the Brooklyn Bridge dominated the lower Manhattan skyline.**

Gustav Lindenthal was born in 1850 to the family of a cabinet-maker in Brünn, a manufacturing city in the Austro-Hungarian province of Moravia, later part of Czechoslovakia. Although he had little education or formal training, he would be hailed as the dean of American bridge engineers.

ried to the United States at roughly the same age—Lindenthal at twenty-six in 1876, Ammann at twenty-five in 1904. The young republic offered enormous opportunity to any able engineer. But it wasn't opportunity alone that lured these two. To a bridge builder, America was a land of wonder.

People who can fly in airplanes have trouble seeing bridges with nineteenth-century eyes. One has to be stuck on the ground, peering at a distant shore with longing or ambition or sheer curiosity, to understand. On one level a bridge was a matter of simple utility—a tool to ease a river crossing by replacing clunky barges and ferries with fast trains. But in America, moving from here to there was not only a practical matter. America's great bridges inspired extraordinary public interest, even reverence. They were symbols of freedom and the great trek west. They were objects of beauty and monuments to industrial prowess, the largest things in the landscape, testifying to the mental and physical might of the builders.

By the Civil War, the technology of transportation had shifted decisively from steamboats to steam locomotives, and the railroads had infected the nation with a bridge-building fever. Suddenly the rivers were not so much arteries of trade as brick walls, and they had to be gotten over. James Eads's career embraced the change. As a river man he built the jetties that saved the Mississippi River for commerce; as a railroad man he spanned the Mississippi with a beautiful arch bridge at St. Louis, completed in 1874. When people of the 1870s said Eads was America's greatest engineer, they were thinking chiefly of the bridge. The architect Louis Sullivan later wrote that, as a boy in Chicago, he had hungered after news of the bridge's construction, for "here was Romance, here . . . was Man, the great adventurer, daring to think, daring to have faith, daring to do." Eads's example inspired other cities and other engineers, and soon rail bridges were spanning rivers all over the East and out into the West, knitting water-divided cities together and connecting distant cities to each other, not to mention the towns and villages along the way. Where old wooden structures stood over rivers, people demanded new spans of iron or steel. The first transcontinental railroad, completed in 1869, was only the central strand in a web of transportation such as the world had never seen, and every strand needed bridges.

Amid this expansion, Gustav Lindenthal became known as a bridge engineer of great skill and superb aesthetic sensibilities. After several years assisting with rail bridges in Chicago, Pittsburgh, and Cleveland, he established an independent practice in Pittsburgh, with its three rivers. He soon had contracts for two major projects. One was a suspension bridge over the Allegheny River. The other, known as the Smithfield Street Bridge, connected downtown to the South Side across the Monongahela. Anchored by ornate, Victorian portal towers, the bridge was a complex, 352-foot form called a lenticular truss, for the lenslike shape of the bowed girders that held up the roadway. It was Lindenthal's misfortune to complete the bridge in 1883, the same year as the Brooklyn Bridge, which was famous as the world's largest bridge even before its completion. Still, the Smithfield Street structure was immediately recognized as one of the country's finest bridges, a "triumph of architectural skill over the gross bulkiness that in the past was considered inseparable from an adequate amount of strength," a critic said.

Lindenthal's Pittsburgh spans constituted a résumé strong enough to attract attention in New York City. Accounts differ about who approached whom, but it's clear that in 1885, Lindenthal held discussions with Samuel Rae, an engineer serving as assistant to the vice president of the Pennsylvania Railroad, about "the practicability of a railroad bridge across the Hudson River."

Smithfield Bridge, Pittsburgh, Pennsylvania, c. 1890. This photograph of Lindenthal's original design shows his affinity for ornate architecture—far more than a bridge needed for strength. The bridge was completed in 1883. In 1915, the ornamentation was simplified.

"IMAGINE . . . A MAGNIFICENT BRIDGE"

To an engineer of Lindenthal's ambition, New York was irresistible, not just because it was the nation's greatest city but because it stood in a maze of waterways—the Arthur Kill and Kill Van Kull, separating Staten Island from New Jersey; the yawning Narrows, where the Atlantic Ocean squeezed between Brooklyn and Staten Island to make the Upper Bay; the East River, sliding up the side of Manhattan; the ribbonlike Harlem River, dividing Manhattan from the Bronx; and the mighty Hudson.

In the 1800s, the city proper was confined to the island of Manhattan, but in the 1880s and 1890s it was obvious to all that its boundaries were likely to expand, and the change promised work for civil engineers. As the city's population and power soared, leaders and civic groups urged the consolidation of all the settled places around the great harbor. To many it seemed a matter of simple destiny to meld these people into a single city—physically divided by water, yes, but commercially united by water, too, like London, Paris, Amsterdam, Prague, and Vienna. Brooklynites, guarding their

"Ferryboats," John C. McCullogh, 1896.
Before New York's bridges were built, ferry traffic clogged the city's waterways and delayed commuters.

city's status as the nation's third largest, were reluctant, but many others were enthusiastic—merchants seeking better docks and city services; property owners with an eye for rising values; leaders in business and the arts, watching the rise of Chicago with jealous unease; and townsmen on the fringes, standing to gain services and development.

A key argument for creating a greater New York was the urgent need to maintain good pathways to the continental interior, a project that would require the financial and political muscle of a united metropolis. By 1885 the Erie Canal was a relic surpassed by rails—mightier than canals because they were cheaper to build and not so dependent on the vagaries of geography and water. Rails could go anywhere. Their only natural enemy was New York's natural ally—waterways, the Hudson River in particular. Trains that had vaulted mountain ranges and crossed endless expanses of desert, prairie, and woodland chugged to a dead stop on the New Jersey side of the Hudson, three thousand feet across and very deep. Freight and passengers had to be laboriously shifted to barges and ferries for a choppy, expensive voyage to crowded docks across the river.

The nation's two greatest railroads—the New York Central and the Pennsylvania Railroad, popularly known as "the Pennsy"—fought a long war over direct access to Manhattan. The New York Central entered the city from the north via a bridge over the narrow Harlem River. The Pennsy, in danger of losing its preeminence as "The Standard Railroad of the World," desperately needed a way across the Hudson, not only to provide direct rail access to Manhattan, but to connect with the New York, New Haven, and Hartford Railroad, thus connecting the Pennsy's routes into New England for the first time. A tunnel was considered, but the problem of smoke from the Pennsy's great locomotives was considered prohibitive. Yet a Hudson bridge would be a staggering proposition. It must leap the river in a single span, and it had to be high enough above the water, even at high tide, to allow the tallest ocean liners and warships to pass underneath. In short, it would have to be the longest, highest suspension bridge in the world.

Here was a challenge commensurate with Gustav Lindenthal's ambition, and he and his railroad allies soon organized a North River Bridge Company to build it. (In maritime circles, the Hudson still went by the name North River, a legacy of colonial times.) The project's sheer size was one attraction. It also offered Lindenthal the chance to compete face-to-face with the Brooklyn Bridge.

If great bridges of the 1800s were regarded as monuments to human ingenuity and power, the Brooklyn Bridge stood as the acme of that industrial pantheon. During the bridge's opening ceremonies

"Hudson River Bridge," Gustav Lindenthal, 1891. **A graphic depiction of Lindenthal's dream of a colossal railway bridge from Manhattan to New Jersey. Note the dwarfed Brooklyn Bridge in the background, at right.**

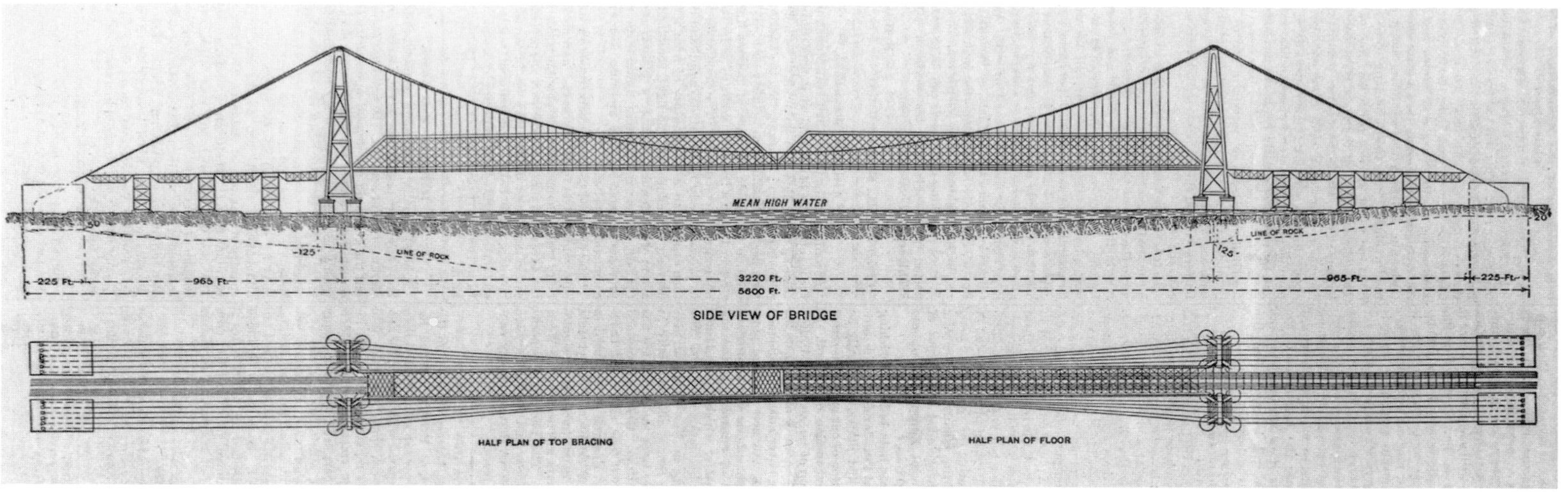

"The Proposed Great Suspension Bridge Over the Hudson River at New York," Gustav Lindenthal, 1891.

in the spring of 1883, a sign in Brooklyn store windows declared: "Babylon had her hanging gardens, Egypt her pyramid, Athens her Acropolis, Rome her Atheneum; so Brooklyn has her bridge," and few who saw the bridge demurred. Designed by the fierce John Roebling (yet another central European bridge builder), who died from an injury suffered at the site, and built by his son, Washington, who was crippled there, the bridge loomed over all other structures in the cityscape of its era. It crossed the East River from lower Manhattan to Brooklyn with a center span of 1,595 feet—by far the world's longest—a roadway wider than Broadway, a promenade for pedestrians, and lovely webs of diagonal cables. Even before it became a magnet for painters, photographers, and poets, the bridge was hailed as the central artifact of its age. The journalist Montgomery Schuyler summed up the enduring view: "It so happens that the work which is likely to be our most durable monument, and to convey some knowledge of us to the most remote posterity, is a work of bare utility; not a shrine, not a fortress, not a palace, but a bridge."

Lindenthal created a design grander—that is, larger—than the Roeblings' in every way, and it grew larger with each revision. Where the Brooklyn Bridge merely united two cities, Lindenthal imagined his bridge as the capstone of the entire transcontinental railroad, with an enormous new Manhattan station at its eastern terminus. In a public proposal, he wrote:

> Imagine . . . in a central part of New York City, within a stone's throw of its greatest avenue, a grand, imposing station, combined with every convenience and comfort of a first-class hotel, with numerous tracks and platforms, accommodating thirty trains at one

> time, arriving and departing, having all the elevated railroads running their trains directly into this station. Then imagine a massive stone viaduct and lofty columns supporting a six-track roadbed, through and over blocks of buildings to a magnificent bridge over the North River, leaping with a single span over its entire width, without a pier or other obstruction and with a clearance above highest tide of 140 ft. . . . Then imagine the . . . tracks continued on a viaduct and gently descending to the level of the country in New Jersey to connections with all existing railroads and for future lines that will be built. . . . Such a project can be realized. It is perfectly feasible and practicable.

He designed it to bear an astonishing load—six lanes of heavy rail traffic, four sets of tracks for lighter passenger trains, roadways for horse-drawn wagons, promenades for pedestrians. The span between towers would be 2,850 feet, nearly twice as long as the Roeblings' span. The walls of the two immense towers would slope skyward to castlelike peaks; these, too, would be taller than the Roeblings'. Indeed, the bridge would all but overwhelm the skyline of west Manhattan, and Lindenthal intended it to do so for a long time. "If well maintained," he said, "there is no reason why it should not last as long as the Egyptian pyramids. . . . No tornado could blow the structure over. No earthquake could shake it down, unless it were so great that the rock would cave and split, and swallow up the North River." In short, Lindenthal suspected that God alone could deprive New York of his bridge.

Lindenthal as New York City bridge commissioner, c. 1902.

He underestimated the power of politics, high finance, and the Army Corps of Engineers. From the mid-1880s to the early 1900s, Lindenthal's plan died one painful death after another. First the corps said the bridge would interfere with naval traffic on the river. Lindenthal shifted the design northward, where the river was narrower and the piers could be placed nearer the banks. But the panic of 1893 took down several of the railroads that were to cooperate with the Pennsy in providing funding. When Lindenthal and the Pennsy tried again in 1898, the remaining rail lines, battered by years of depression, said no.

The Pennsy gave up and commissioned a tunnel and a different terminal; thus the demise of Lindenthal's plan in the 1890s set the stage for the mighty Penn Station, which opened in 1911 as the Pennsy's Manhattan terminus. Reluctantly, Lindenthal set aside his grand design for connecting New York with the rest of the country and turned to the business of stitching together the boroughs of Greater New York. He became commissioner of the city's department of bridges. In that role he helped to launch a massive campaign of bridge and tunnel building that would transform the region from an archipelago to a seamless metropolis. Lindenthal completed construction of the Williamsburg Bridge and oversaw a revision of the design for the Manhattan Bridge. (The original design for the latter, he said, had been "the ugliest possible.") His signature work in this period was a massive cantilever bridge to cross the East River to Queens—the Queensboro Bridge, often called the Fifty-ninth Street Bridge. After only two years in the city's employ, Lindenthal reestablished his private practice, chiefly as a consultant to the railroads. All the while, he and his old friend Samuel Rae continued to talk of a bridge over the Hudson. Someday, they believed, circumstances would break their way and the bridge would rise, and many others assumed they were right.

By now, with the Queensboro to his credit, Lindenthal at fifty-five was the acknowledged leader of his profession and a prominent man-about-town. With his third wife and their infant daughter, he

The Queensboro Bridge, c. 1910. The cantilever principle behind Lindenthal's Queensboro Bridge, which had been used for centuries in wooden bridges, gained new attention in the 1890s in the designs of James Eads and John Roebling. A cantilever bridge is built out piece by piece from supporting piers. Then the center sections that complete the span are raised from barges or lowered into place by cranes. The bridge is anchored on the piers by counterbalancing elements.

lived on a New Jersey estate which he named The Lindens. He took a prominent role in the Liederkranz, the leading German-American club New York, and served as president of the Austrian-American Society. Railroads continued to seek his guidance. He worked in a skyscraper office in the financial district, surveying the city's waterways like a lord his fiefdom.

In 1904, the Pennsy asked Lindenthal to develop a plan for fulfilling the railroad's intention to link New Jersey with New England. It was to be a relatively short but highly complex affair called the New York Connecting Railroad. The line would start at the rail yard at Sunnyside, Queens, rise over Astoria to cross the turbulent section of the East River called the Hell Gate, proceed by a raised viaduct over Ward's Island and Randall's Island, then hop the Bronx Kill to join up with the New Haven Railroad in the South Bronx. (The route runs roughly parallel to the later Triborough Bridge complex, minus the Triborough's main bridge into Manhattan.) A great bridge was needed at the center of this new connecting line, over the Hell Gate. The Pennsy wanted the Hell Gate Bridge to be yet another monumental structure. "Penn Station was created to be its Pan-

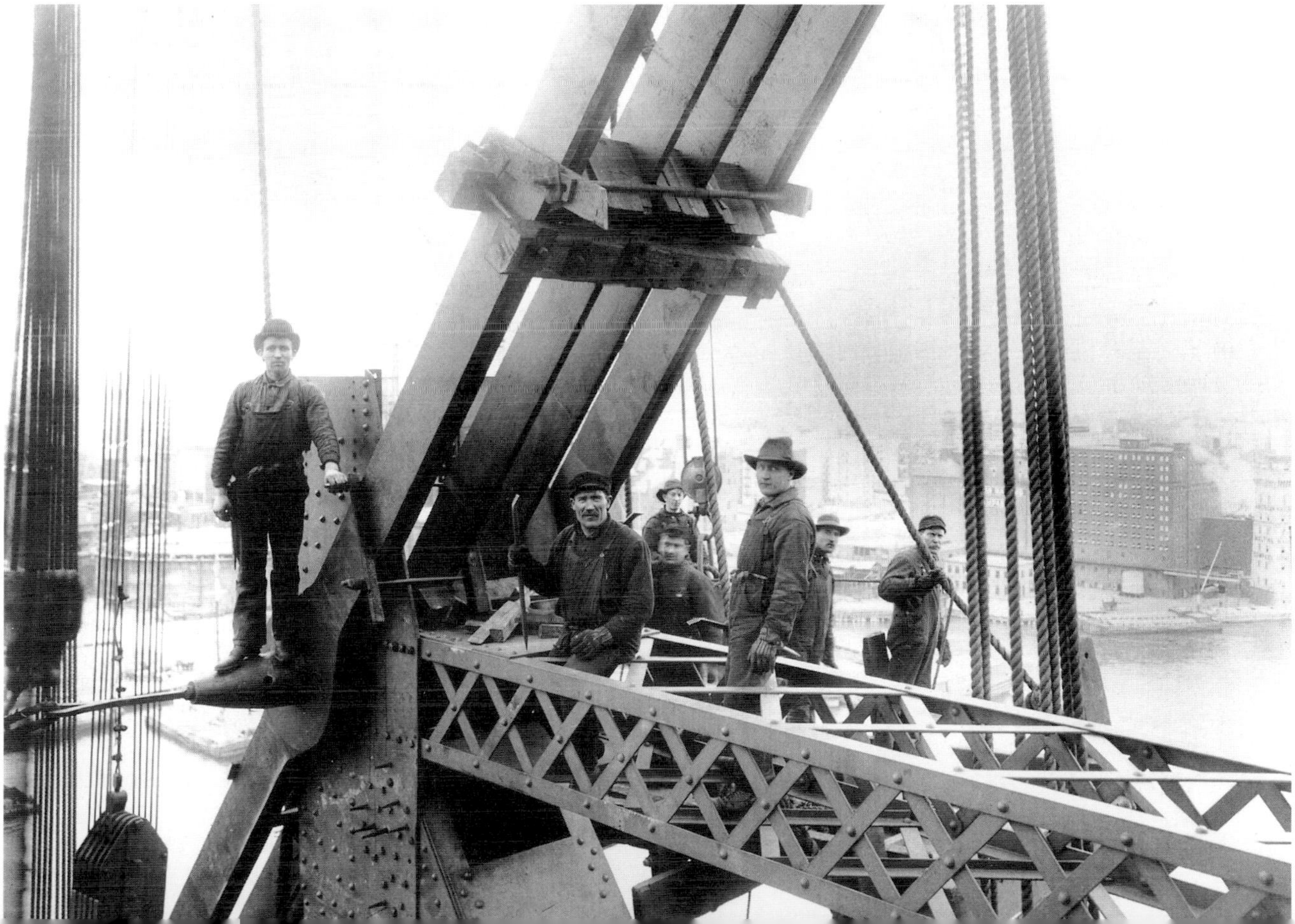

Workers on Queensboro Bridge, c. 1908. Although construction of the Queensboro began after Lindenthal left as bridge commissioner, the bridge was built in keeping with his designs. A steel strike in 1905 delayed construction, but the project was begun in earnest in 1906 and hailed as the largest cantilever bridge in the United States.

theon," the writer Tom Buckley said, "and the Hell Gate was to be its triumphal arch."

In 1907, Lindenthal began work on the design. It was not to be a suspension bridge but a single-span arch—the world's longest. To build it, he would need help. In 1910, at a party in Philadelphia, he met a highly promising young engineer—Othmar Ammann.

"FOR ONE THOUSAND YEARS . . ."

Ammann was one of those people whose path toward their life's work seems enviably short and straight. Growing up in Switzerland, he hoped at first to become an architect. But at the state college in Zurich he showed brilliance in mathematics, so he shifted to a course in engineering at the prestigious Swiss Federal Polytechnic Institute. One of his professors there had built bridges for the Northern Pacific Railroad. He described the American West to his students and showed them photographs of the great American bridges. There was so much to be built in America, he said, that a young engineer could build things by the age of thirty that few European engineers could hope to achieve in an entire career. So Ammann decided to visit the United States for a couple of years, see the World's Fair at St. Louis,

Queensboro span is completed, March 12, 1908. As politicians and engineers looked on, the last link in the Queensboro was lowered into place, creating a bridge that "seemed to be defying the law of gravity." A debate ensued over the bridge because a cantilever bridge had collapsed in Quebec the year before. *Scientific American* aggravated concerns when it reported that changes in Lindenthal's design would upset the balance. Two independent consultants examined the bridge and proclaimed it perfectly sound. In August, construction crews connected the bridge to its approaches, and pedestrians began walking the mile span between Queens and Manhattan.

Opening Day, Queensboro Bridge, 1909. The bridge opened on April 11, 1909. Although "Queensboro" was accepted in the end, there was controversy about the name, including an effort by Irish-Americans to eradicate from the United States all place names of English origin.

get some experience, and return to a career in Switzerland.

But America seized him in an unbreakable embrace. With his respected Swiss degree, he got good jobs immediately, and the work delighted him. In the summer of 1905 he rushed home to marry Lilly Wehrli, a girl he had loved since their childhood days along the Rhine, then steamed back after a quick honeymoon in Italy. The Ammanns assured themselves and their families they would come home for good soon. But each time they made plans, Othmar's latest success intervened. "The more I learn the more ambitious I become and the more I enjoy my work," he wrote his parents.

Aiming at a career in long-span steel bridges—the most visible and prestigious structures in all of engineering, offering full-time employment to only a small elite—he apprenticed himself to one long-span authority after another: Joseph Mayer in New York; Ralph Modjeski in Chicago; Frederic Kunz in Philadelphia. Each tried to keep Ammann; each gave him his blessing and a glowing recommendation when the quiet Swiss moved on. In Harrisburg, Pennsylvania, where the Pennsylvania Steel Company had five thousand men building bridges, Ammann's supervisors took note of the youngster who spent his lunch hours prowling the factory, studying

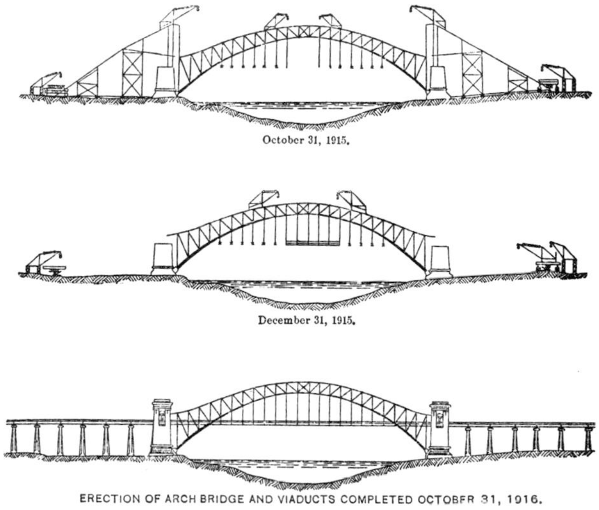

Plans and progress of Hell Gate Bridge construction, 1916.

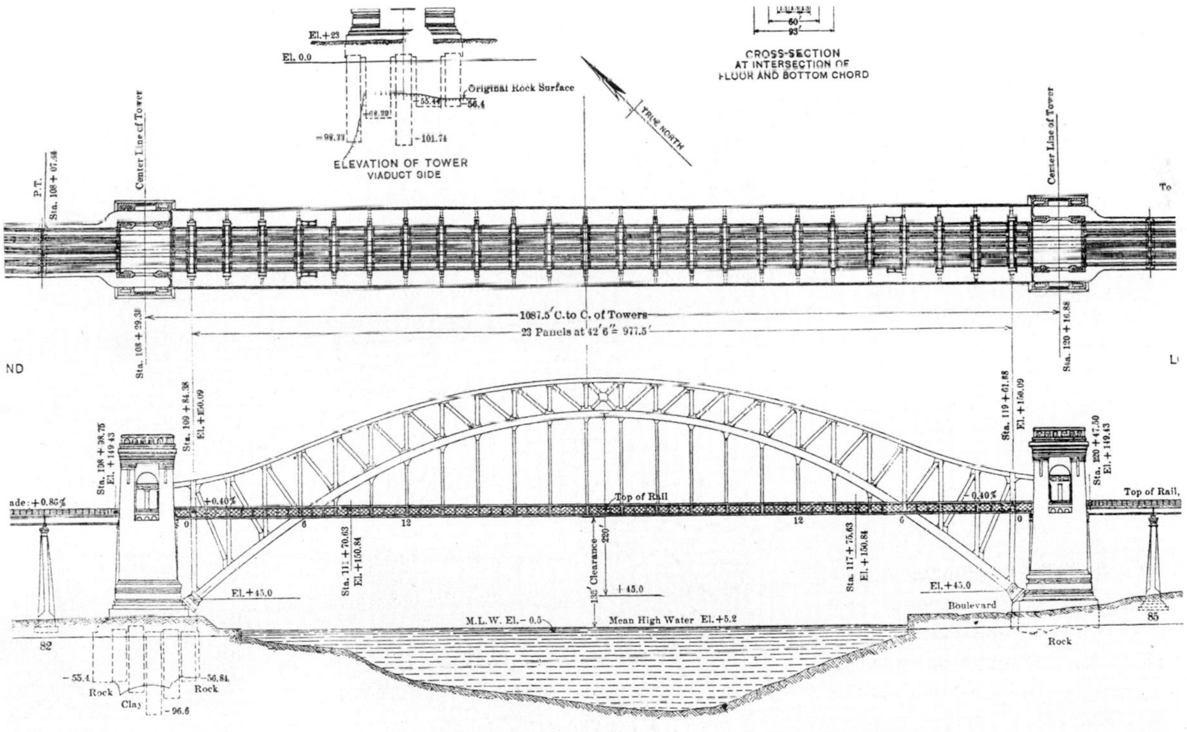

Hell Gate Bridge, September 30, 1915. Lindenthal considered three designs for the Hell Gate–cantilever, truss, and suspension–all with an 850-foot main span. He chose the aesthetically pleasing arch. These photographs show the center piece of the arch being raised into place.

Hell Gate Bridge, c. 1917. Upon its completion in 1917, it was the longest steel arch in the world. Its ornamental towers play no structural role.

workmen as they forged steel parts and put things together. Soon they were all asking *him* for help with nitty-gritty problems of construction technology. When the cantilever Quebec Bridge collapsed into the St. Lawrence River in 1907, killing seventy people, Ammann went immediately to Quebec, where he became a key figure in the inquiry and helped with a new design. By his thirtieth birthday he had played a role in dozens of bridges in several states.

In 1910, the Ammanns again were preparing to move home. But then, through Frederic Kunz, they met Gustav Lindenthal. Despite an age difference of nearly thirty years, the two men became good friends and saw each other often. When Lindenthal offered Ammann the job of first assistant on the Hell Gate Bridge project, with power over all construction and ninety-five subordinate engineers, the Ammanns' plans for a return to Europe once again were set aside.

The two men made a powerful team of opposites. Lindenthal's daughter remembered that "you always knew when my father entered a room, but Ammann was modest and self-effacing." Lindenthal looked like his bridges—tall, thick, and dignified, with Victorian ornamentation, in his case a mustache and beard that he wore from early manhood until the end of his long life. Ammann was short, slender, and normally clean-shaven. "Lindenthal . . . was the mercurial, impulsive genius, full of Wagnerian ambitions and vainglorious dreams," a journalist wrote later. "Ammann was the quiet, almost colorless Swiss burgher, tactful and courteous in his re-

lationships with other men, calm in his judgments, flawless in his engineering, and infinitely abler as an executive."

Together, their talents built an impressive and lasting bridge—but the Hell Gate design was Lindenthal all over. As Tom Buckley put it later, "He believed that a bridge, and particularly a railroad bridge, should not only be strong but look strong." It was not enough to create a simple, graceful steel arch. With help from an architect, Lindenthal placed imposing masonry towers at the anchorages. With their arches and cornices, the towers looked like turrets in a castle. Yet they played no role at all in holding up the bridge. Tear down the towers and the Hell Gate Bridge would carry trains as safely as ever. The towers are there because Lindenthal thought them noble.

Lindenthal (center, with beard) and Ammann (seventh from left) visit the Hell Gate Bridge during construction, c. 1917. The two engineers, thirty years apart in age, enjoyed a close rapport.

Once he became Lindenthal's dutiful, indispensable, and unheralded lieutenant, Ammann paused in his upward march through his profession. But he offered no sign that he felt stymied or resentful, or that he itched to draw and build his own designs. On the contrary, he seemed to feel privileged. When he wrote the definitive report on the Hell Gate Bridge, he praised Lindenthal's conception as "one of the finest creations of engineering art of great size which this century has produced."

In return for Ammann's friendship and help, Lindenthal helped his protégé. When war broke out in Europe in the summer of 1914, the younger engineer, still a Swiss citizen, dismayed his master by rushing home for army service. But when the German army bypassed neutral Switzerland and Ammann returned to New York, Lindenthal immediately welcomed him back to his old job; indeed, the older man had tried to intercede with the Ammann family to procure Othmar's early release from service. They finished the Hell Gate together in 1917 as well as an enormous bridge to carry coal trains across the Ohio River near Cincinnati. When America's entry into the war dried up the stream of bridge contracts, endangering Ammann's livelihood, Lindenthal steered him to a job in New Jersey, where Ammann would run a clay-mining company in which he, Lindenthal, held a financial interest, with the understanding that once the war ended, they would build bridges again.

Now approaching his forties, Ammann must have suspected that fate was laughing at him—a digger of holes in the ground who wanted only to raise towers and cables toward the sky. Every day he labored dutifully down in South Amboy, but "the position was not attractive," he allowed himself to admit. He accepted it, he said, "to be on hand in case Mr. Lindenthal needed my assistance."

Yet, during this gloomy interval, a series of events began that would transform Ammann's career and the face of New York City. Little of this drama was known until Jameson Doig, a political scientist at Princeton University, discovered a trail of forgotten letters and documents and followed them to the full story, which he has told in two scholarly articles and a book, *Empire on the Hudson.*

It started when Ammann's talents as an industrial manager turned the shaky Such Clay Pottery Company into a profit maker. This inspired the admiration of an influential member of the company's board, George Silzer, a Democratic politician. From 1909 to 1913, Silzer had served as chief aide to Woodrow Wilson, then governor of New Jersey. Now Wilson was president, and Silzer was moving toward his own run for governor. He would prove a good man to have impressed when, after the war, people again demanded a better way to cross the Hudson than by barge or boat.

No one who knew Gustav Lindenthal was surprised to learn he was more than ready to answer this call, though he was now seventy years old. He and Samuel Rae had kept the North River Bridge Company alive for twenty years, waiting for just such a rise in fortune, and Lindenthal had made his plan grander than ever. Unveiled in increasingly larger versions from 1920 to 1923, the new design called for steel and granite towers higher than the Woolworth Building, then the tallest in the world. The span between towers would be nearly half a mile long. There seemed no limit to the traffic Lindenthal said the bridge would bear: four freight trains at a time, four heavy passenger trains, light trolleys, automobiles, trucks, and pedestrians speeding across the bridge on conveyor-belt sidewalks—a half-million people per hour, Lindenthal claimed. One cross-section view of the project envisioned space for twenty-eight lanes of vehicular traffic—or was it thirty-two lanes?—not to mention plenty of space for families on their Sunday stroll. The location was to be north of the earlier versions. From Weehawken, the bridge would cross the river to Fifty-seventh Street, soaring down over many city blocks to its terminus under an enormous office building smack in the middle of midtown, already packed with buildings and traffic.

The city was accustomed to monumental building projects. Still, the magnitude of what Lindenthal wanted to do stunned even New Yorkers. The engineer took the offensive, saying the bridge would be more economical than the multiple tunnels required to serve comparable traffic. But he was never one to base an argument on sheer practicality. He appealed to the judgment of posterity: "Our city will be pre-eminently the city of great bridges, representing emphatically for centuries to come the civilization of our age, the age of iron and steel. A time must come, not many generations distant, perhaps not more distant than the crusades in the past, when the building of such colossal structures will cease because the principal material of which they are molded, that is, iron and steel, will not be longer obtainable in sufficient quantity and cheapness. When the iron age has gone, the great steel bridges of New York will be looked upon as even greater monuments than they are now."

In 1920, as this debate was heating up, Lindenthal called Ammann back to work, promising an eventual partnership in a new firm he planned to found. It was a gamble for Ammann. Money was still tight in Lindenthal's postwar office, and a great deal depended on winning approval of the Hudson project. Ammann agreed to come back, though this time as a consulting engineer, not a salaried employee. For two years he divided his time between three bridges in Oregon and preparations for the staggering Hudson project. In the early days he was enthusiastic. "The new project brings me great satisfaction," he wrote his mother. "It is a great noble structure, and . . . the concept and modeling of the project demand intense attention and work."

1921. Lindenthal's massive Hudson River bridge design in comparison to the Brooklyn Bridge, completed four decades earlier.

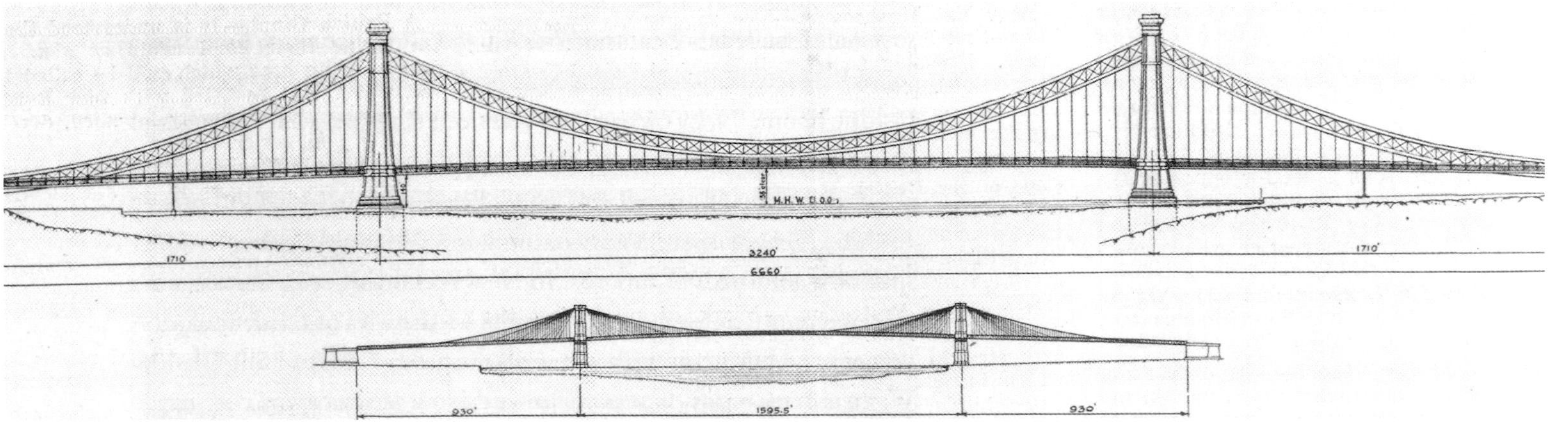

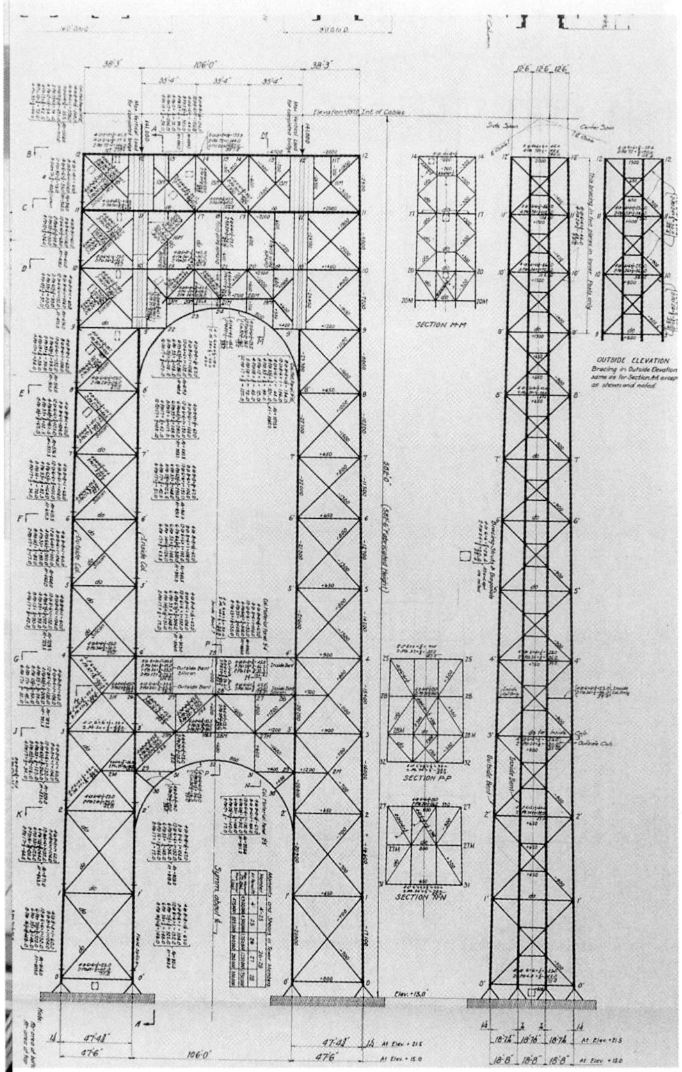

Ammann's design for the towers' internal structure. He intended that the steel skeleton would disappear under concrete skin. But it never did.

left as stark steel skeletons, unlike any other bridge towers ever built. By the time the depression ended, the towers had grown on people, and the idea of the facings was dropped for good. Even the chief engineer had grown to accept his unfinished offspring as permanent. In his final report on the construction, Ammann said it would be nice to add the facings someday. But the exposed steel "twigs," he said, "lend the entire structure a much more satisfactory appearance than [I] (and perhaps anyone connected with the design), had anticipated."

The Cables

Ammann could choose between two equally effective systems for suspending the roadway: chains of steel members called eyebars, or cables to be spun on the site. The choice was to be based on economy. He invited bids and awarded the contract to a firm with a familiar name: John A. Roebling & Sons of Trenton, New Jersey, which had been making bridge cables since before the firm's founder designed the Brooklyn Bridge.

When the towers were complete, a guide cable was hauled from one side of the river to the other and hoisted atop the towers in the familiar inverted-arch shape. Then a wheel trailing a thin steel wire was sent up the guide wire and over the towers; then back and forth, day after day. Four hundred and thirty-four of these wires made a single large strand. When 61 strands were complete, they were bound together with wire and compacted to form a single cable 3-feet thick. The bridge had 2 pairs of these cables—107,000 miles of steel wire in all, weighing 28,100 tons. It took 10 months to spin them.

The Roadway

When the time came to construct the deck, Ammann continued to believe he could do without what he called "clumsy stiffening trusses." He did, however, add a more subtle stiffening device: plate girders built into the underside of the roadway, a kind of horizontal truss that added little to the deck's thickness. Thus, the deck that emerged over the Hudson River was only twelve-feet deep. From any distance it looked like a long strip of cardboard, so light that it appeared to float, not to hang, at the ends of its suspender cables. Yet when winds came up and the engineers watched, the roadway stayed stiff and unmoving. Ammann's application of the deflection theory had worked. The roadway had space for six lanes of auto traffic and two promenades for pedestrians. Ammann expected auto traffic to grow, so he left a thirty-two-foot-wide path in the center to accommodate two more lanes should they become necessary later.

The Result

In the end, the bridge was completed $1 million under budget and eight months ahead of schedule. Cars and trucks first moved out over the ribbonlike deck on October 25, 1931. The bridge vindicated Ammann's vision of an automotive future even as it helped to bring that future into being. In 1932, the first full year of operation, the bridge carried some 5 million vehicles. In 1946, rising traffic called for the addition of two more lanes of traffic in the space Ammann had left for just that purpose. By the end of the twentieth cen-

Opening Day, George Washington Bridge, October 24, 1931. Even in the midst of the Great Depression, 55,523 cars passed over the bridge on Opening Day.

tury it was carrying 50 million vehicles per year, making it one of the world's busiest bridges. In time, as other bridges and arteries were built, it became the key structure in a web of ground transportation that links the rest of the nation with New York City, Long Island, and New England.

The safety of Ammann's extraordinary design was reexamined in the wake of the most spectacular bridge failure in American history—the collapse of the Tacoma Narrows Bridge over Puget Sound in Washington State, on November 7, 1940. Only a few months after the bridge was finished, wind in the sound set off a terrifying ripple in the roadway. As newsreel cameras whirred, the ripple became a rolling wave, and the span gave way and crashed into the water below. The bridge's design was the work of Leon Moisseiff, prophet of the deflection theory, whose reputation was ruined by the disaster. Under Ammann's leadership, a congressional inquiry concluded that Moisseiff had unintentionally lent his roadway a shape that bore a lethal resemblance to an airplane wing. When confronted with even a relatively mild wind, the roadway wanted to fly. Ammann's bridge over the Hudson, much heavier than the Tacoma bridge, never behaved this way, and Ammann believed it was unnecessary to add trusses for strength. But by the late 1950s, traffic on the George Washington Bridge had reached its maximum, and a lower deck was added. With its trusswork connections to the original deck, the new deck only increased the bridge's power to resist oscillation. The extraordinary lightness of the original appearance was lost, though the span's appearance remains delicate compared to many bridges. The new deck opened to traffic in 1962.

The bridge was named for the nation's first president, who led patriot troops against the British at the site. But among New Yorkers it became known simply as "the G.W." or even "the George," as if it were a member of the family. It has achieved a reputation as one of the world's most beautiful large structures, despite the strange appearance of the unfinished towers. Le Corbusier, the French architect who wielded enormous influence over the emerging international style, crossed the bridge in a car and delivered a verdict in which many have joined, from artists and architects to New York commuters:

"The George Washington Bridge over the Hudson River is the most beautiful bridge in the world. Made of cables and steel beams, it gleams in the sky like a reversed arch . . . It is the only seat of grace in the disordered city. It is painted an alumninum color, and between water and sky, you see nothing but the bent chord supported by two steel towers. When your car moves up the ramp the two towers rise so high that it brings you happiness; their structure is so pure, so resolute, so regular, that here, finally, steel architecture seems to laugh."

David P. Billington, a professor of civil engineering at Princeton University and a leading historian of technology, has hailed the bridge as a key example in the modern tradition he calls structural art, a form that blends the techniques of heavy industry with the aesthetic sense of a sculptor. The form emerged in the 1800s, most spectacularly with the Brooklyn Bridge and the Eiffel Tower. Structural art blends "efficiency, economy, and elegance," Billington contends. It is not art for art's sake; it serves a society's utilitarian needs. And it must conform to economic and political constraints. Practitioners reject the *beaux-arts* tradition that Gustav Lindenthal represented, with its insistence on decorative flourishes and its monumental mentality. Nonetheless, structural art reflects the aesthetic choices of the designer. Together, engineering and artistry produce useful structures that call forth a sense of wonder and aesthetic appreciation in the onlooker. Billington believes structural art is a distinctively democratic tradition: "These forms imply a democratic rather than an autocratic life. When structure and form are one, the result is a lightness, even a fragility, which closely parallels the essence of a free and open society."

Lighthouse under the George Washington Bridge, c. 1931. *The Little Red Lighthouse and the Great Gray Bridge* (Harcourt Brace and Company, 1942), a children's book by Hildegarde H. Swift and Lynd Wardin, portrays the conflict between an old lighthouse and the newer bridge. "'Am I brother of yours, bridge?' asks the lighthouse. 'Your light was so bright that I thought mine was needed no more.' 'I call to airplanes,' cried the bridge. 'I flash to the ships of the air. But you are still master of the river. Quick, let your light shine again. Each to his own place, little brother!'"

With the George Washington Bridge, Billington wrote in his seminal study, *The Tower and the Bridge,* Othmar Ammann emerged as a central figure in the structural art tradition, abandoning the heavy masonry adored by his mentor, Lindenthal, and insisting on a new aesthetic of lightness. His later work would bring the tradition to fulfillment, and thus make structural art perhaps the defining characteristic of the landscape that Americans built.

Details likely will never be known, but it's clear from Ammann's writings that at some point during the bridge's construction he did Gustav Lindenthal the courtesy of consulting him on design ques-

Triborough Bridge construction, July 24, 1935. The Hell Gate Bridge rises behind the construction site of Ammann's three-way bridge, which links Manhattan, Queens, and the Bronx.

projects around the world, including the Golden Gate Bridge.

Throughout his career, like all engineers, Ammann was commissioned to design structures that never were built. He always took it in stride. Among the lessons he had learned from Lindenthal was that politics and finance matter as much to the completion of a bridge as engineering. But there was one unbuilt span he must have regretted. In 1963 the Swiss Federal Engineering Department asked him to design a suspension bridge to relieve congestion in the city of Geneva. But final authorization for the Pont sur la Rade was never given.

Just a few years earlier, as Ammann approached his eightieth birthday in the late 1950s, he began planning for what would become—for the second time in his career—the world's longest suspension bridge. This was the Verrazano-Narrows, linking Brooklyn and Staten Island, a structure so large that its design had to take heed of the curvature of the earth. It would be finished the way every one of Ammann's bridges was finished—under budget and on time.

In the years of its construction, speaking with Othmar Ammann became a kind of communion with history, at least among those who appreciated the art of bridge engineering. In 1963, Francesca Gebhardt, the daughter of Gustav Lindenthal, telephoned him. "He invited me to come to his office," she recalled. "He could not have been more considerate. I knew he was very ill, and I was deeply affected. He told me how much he had respected my father and how much he had learned from him. Neither of us alluded to the unpleasantness that had taken place."

In 1964, the writer Gay Talese, then a reporter on the *New York Times*'s metropolitan desk, went to see Ammann for an article to mark the engineer's eighty-fifth birthday. The Ammanns greeted Talese in their apartment on the thirty-second floor of the Carlyle Hotel, on Manhattan's Upper East Side. Ammann escorted Talese from window to window, pointing to each of his New York bridges—the George Washington; the Bayonne; the Bronx-Whitestone; the Throgs Neck; the Triborough; the ingenious little pedestrian bridge that Ammann had constructed over the Harlem River; and the unfinished Verrazano-Narrows, its cranes piercing the horizon twelve miles to the south. Ammann kept a telescope handy to check on the progress of construction.

Talese asked Ammann which was his favorite. Ammann said the George Washington. Whenever he and his wife passed it, he said,

BRIDGEMEN IN THE SKY

In the 1960s, the writer Gay Talese described the men who built Othmar Ammann's last and longest bridge, the Verrazano-Narrows, in The Bridge *(1964). The riveting process went on as it had thirty years earlier, on the George Washington Bridge. An excerpt:*

Building a bridge is like combat; the language is of the barracks, and the men are organized along the lines of the noncommissioned officers' caste. At the very bottom, comparable to the Army recruit, are the apprentices—called "punks." They climb catwalks with buckets of bolts, learn through observation and turns on the tools, occasionally are sent down for coffee and water, seldom hear thanks. Within two or three years, most punks have become full-fledged bridgemen, qualified to heat, catch, or drive rivets; to raise, weld, or connect steel—but it is the last job, connecting the steel, that most captures their fancy. The steel connectors stand highest on the bridge, their sweat taking minutes to hit the ground, and when the derricks hoist up new steel, the connectors reach out and grab it with their hands, swing it into position, bang it with bolts and mallets, link it temporarily to the steel already in place, and leave the rest to the riveting gangs.

Connecting steel is the closest thing to aerial art, except the men must build a new sky stage for each show, and that is what makes it so dangerous—that and the fact that young connectors sometimes like to grandstand a bit, like to show the old men how it is done, and so they sometimes swing on the cables too much, or stand on unconnected steel, or run across narrow beams on windy days instead of straddling as they should—and sometimes they get so daring they die.

Once the steel is in place, riveting gangs move in to make it permanent. The fast, four-man riveting gangs are wondrous to watch. They toss rivets around as gracefully as infielders, driving in more than a thousand a day, each man knowing the others' moves, some having traveled together for years as a team. One man is called the "heater," and he sweats on the bridge all day over a kind of barbecue pit of flaming coal, cooking rivets until they are red—but not so red that they will buckle or blister. The heater must be a good cook, a chef, must think he is cooking sausages not rivets, because the other three men in the riveting gang are very particular people.

Once the rivet is red, but not too red, the heater tong-tosses it fifty, or sixty, or seventy feet, a perfect strike to the "catcher," who snares it out of the air with his metal mitt. Then the catcher relays the rivet to the third man, who stands nearby and is called the "bucker-up"—and who, with a long cylindrical tool named after the anatomical pride of a stud horse, bucks the rivet into the prescribed hole and holds it there while the fourth man, the riveter, moves in from the other side and begins to rattle his gun against the rivet's front end until the soft tip of the rivet has been flattened and made round and full against the hole. When the rivet cools, it is as permanent as the bridge itself.

Verrazano-Narrows Bridge construction, September 25, 1963.

they bowed and saluted. Then Talese asked him to describe his feelings about the bridge.

"It is as if you have a beautiful daughter," he replied, "and you're the father."

In fact, Ammann did have a beautiful daughter. She was born in 1922, about the time her father began to sketch his plans to bridge the Hudson River. She grew up to become an obstetrician. In retirement, Dr. Margot Ammann Durrer was asked for *her* feelings about her father's bridges. She replied: "Perhaps the one I felt in competition with was the George Washington Bridge. As an adolescent, you feel a little put back when people are more interested in what your father's doing than what you are doing. So [at] the mention of the George Washington Bridge, I always said, 'That is my older sister.'

"Now, as an old lady, I think the George Washington Bridge would be my favorite, because I realize that was his firstborn, and it was a rather difficult delivery. Being an obstetrician, that says something to me."

Ammann and Lyndon Johnson, 1964. Several months after the opening of the Verrazano-Narrows Bridge, Ammann was selected to receive the National Award of Science. Margot Ammann Durrer would later recall, "We were playing cards at the time, and after Father heard the news, he calmly said, 'Whose turn is next?' So we went on with the game; Mother and I were so excited that Father easily won. He then went to bed for a peaceful sleep, while Mother and I lay awake for hours, wondering what we would wear for the ceremony."

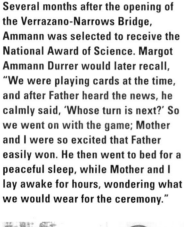

Othmar Ammann and his daughter Margot in 1924.

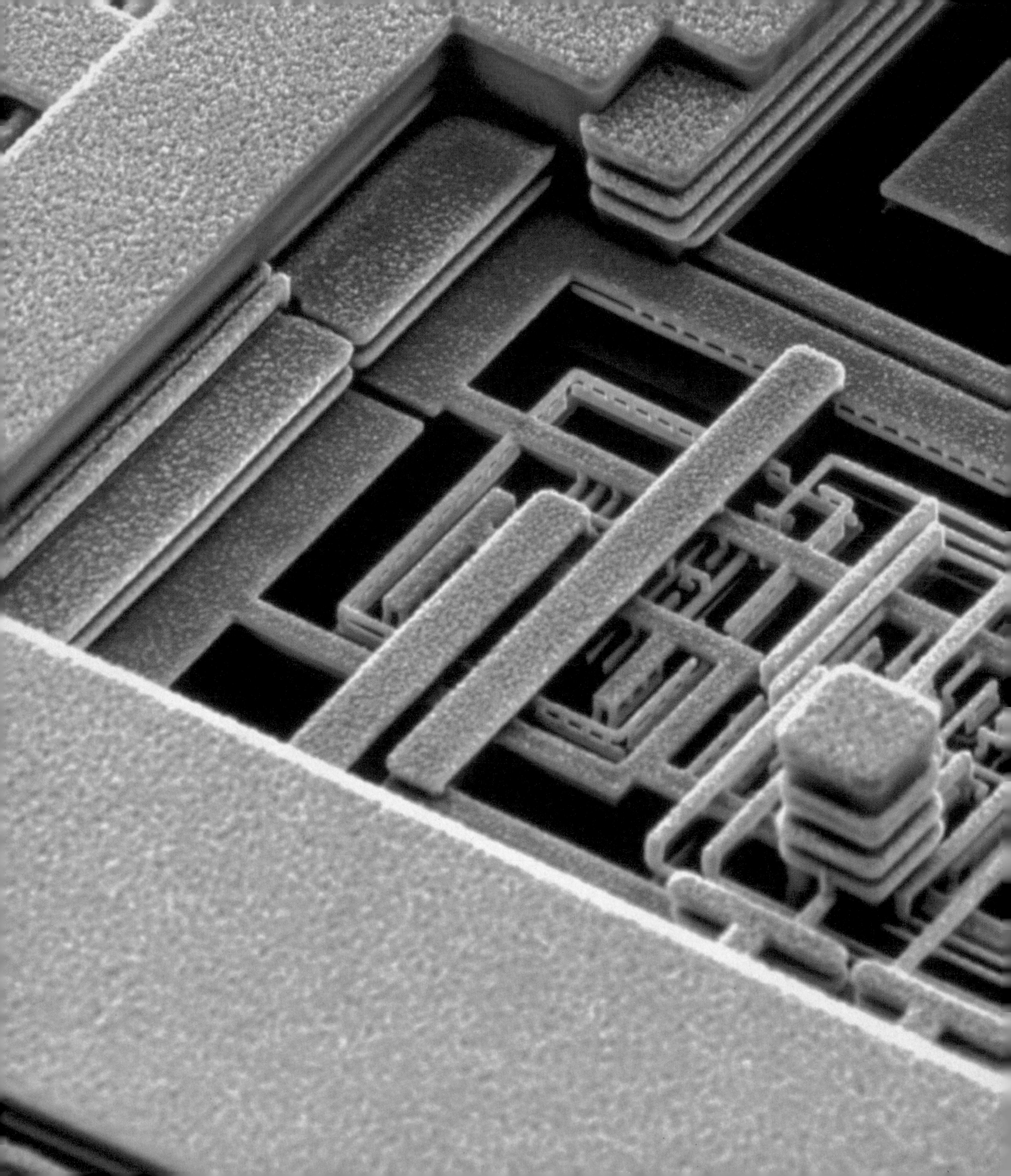

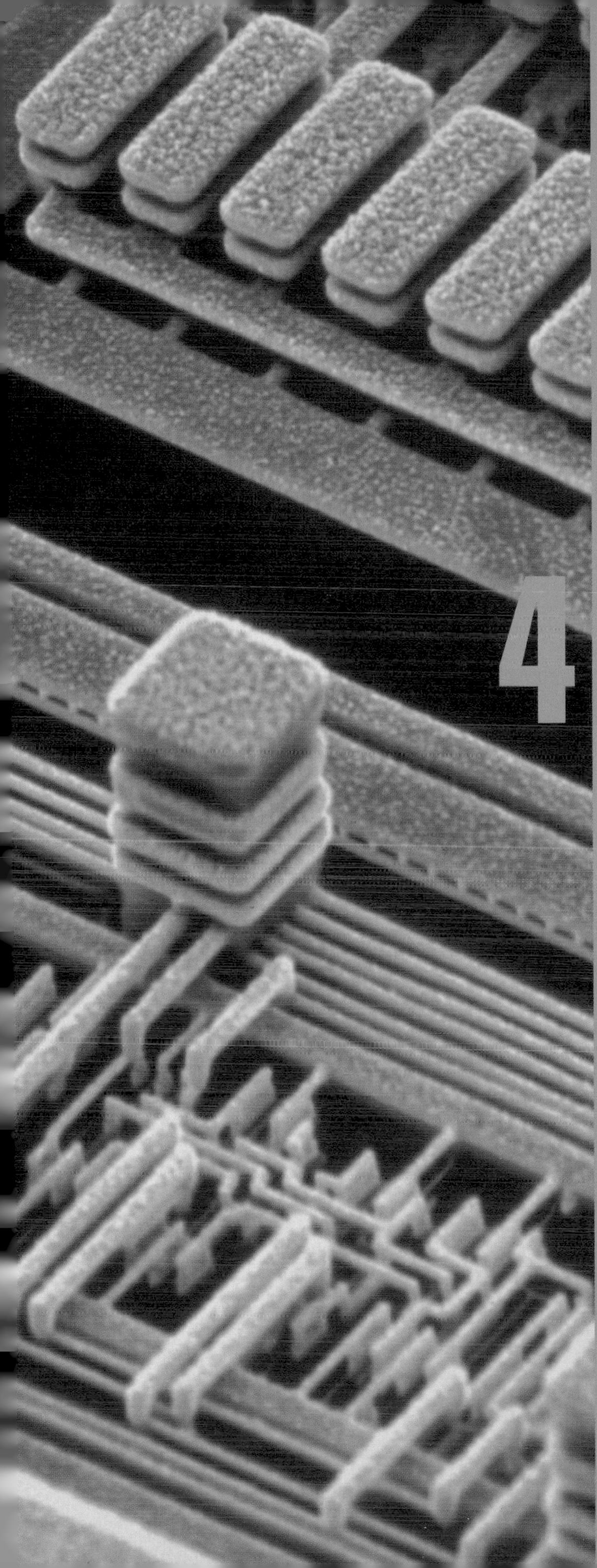

4 PATHWAYS

"Every network was undertaken for its own workaday reasons, to solve some practical problem. Yet each network exerted its own profound influence on the texture of American life, binding Americans together in ever more intricate and intimate webs."

The nation's history might be written solely as a story of pathways, of networks laid one upon another as the generations passed. Their design and construction was the work of engineers who found that pathway building put them at the center of national action, whatever the era. Every network was undertaken for its own workaday reasons, to solve some practical problem. Yet each network exerted its own profound influence on the texture of American life, binding Americans together in ever more intricate and intimate webs.

In so vast a landscape, overland transportation has sorely taxed American ingenuity and resources since colonial times. Though the horse and wagon loom large in the popular memory of early America, the fact is that early roads were no good for any long haul. Natural waterways and man-made canals, not roads, were the networks by which the land was settled. The exceptions were the great overland trails of the West. But these, like the rivers and canals, soon were superseded by the railroads, which reduced even the longest distances to a matter of days and went where rivers and canals could not go.

Yet roads would have their day. It arrived on the heels of the Model T, an automobile that Henry Ford intended mainly as a help to farmers. But the affordable automobile would not be just a wagon with a motor. It so perfectly suited the American urge to get up and go that it spread in a great craze to city and suburb. Soon urban Americans were demanding good roads, not just in town but out in the country—safe, paved roads where they might satisfy their yearning to see beyond the horizons, and not just according to the predetermined routes and schedules of the railroads, but on their own whim and in any direction. Their demands led in time to the most ambitious road network ever conceived—the interstate highway system, nearly half a century in the making.

But the overwhelming popularity of the automobile and the interstates bred large problems. Congestion, pollution, and sprawl drove many people to question the very ideal of technological progress. In the 1960s, the conflict between roads and their critics was magnified in Boston, where the mistakes and excesses of highways builders provoked an open rebellion. What happened next was an extraordinary exercise—first in civic debate, then in civil engineering. In a long struggle that raised new controversies—from the role of the federal government in local projects to the enormous cost of great public works—Boston and its neighbors rethought the proper relationship between people and their pathways. Then they undertook an enormous effort to correct technological folly with technological wisdom.

In the nineteenth century Americans pioneered a new sort of network—the wires that allowed instantaneous communication over long distances. Daniel Boorstin has said that of all the distinctions erased by American life, one of the most striking was this new wiping out of the distinction between "here" and "there." It began with the telegraph system, which allowed coded messages to be transmitted from station to station. Next came Alexander Graham Bell's telephone. Here the link was person to person, not by code but by the natural medium of speech. The harnessing of an invisible entity called the electromagnetic spectrum allowed speech, then images, to be sent through the air via radio and television.

The next advance—surely not the last—rose up on the foundation of all these innovations and one more, the personal computer. Even more than the interstate highways, which put a premium on freedom of movement, the Internet expressed the democratic tendencies of technology in America. It began as a set of tools for engineers and scientists. But the tools escaped from the labs, and ordinary people took them into their own hands. They used them to fashion a constellation of human connections that even Edison could never have imagined.

CHAPTER SEVEN

THE BIG DIG

In my book, the Panama Canal was an easy job.

—*Lou Silano, engineering manager, Boston Central Artery/Tunnel Project*

When Fred Salvucci was a kid in the Brighton neighborhood of Boston in the 1940s and '50s, the grown-ups in his house talked construction all the time. They always had. Fred's great-grandfather on his mother's side had left Italy to work on South Street Station, the biggest rail depot in the world when it opened in 1900. His grandfather on his father's side helped build the dam that gave twentieth-century Boston its drinking water. His dad was a master bricklayer and a major masonry contractor. Uncles, cousins, brothers—everybody worked construction. So it was natural for Fred to follow the family trade. He did especially well in school, well enough to get into the Massachusetts Institute of Technology, the best engineering school anywhere. A big deal for any family, but especially for the Salvuccis, because they always had been builders. So Fred was going to become a civil engineer just when highways and bridges were going up all over the country in the biggest surge of public construction ever seen.

In MIT classrooms on the Cambridge side of the Charles River, he began to learn big-picture stuff about superhighways and urban grids—how to create whole landscapes of urban infrastructure. At night, back in the Salvuccis' three-decker over in Brighton, he would hear the same things discussed. Only now the Salvuccis weren't talking about what the big projects were doing *for* the family—it was what they were doing *to* the family. One of Fred's friends lost his house in the West End to a big urban renewal project. His grandmother, a widow of seventy, lost her home in North Brighton to the Boston extension of the Massachusetts Turnpike. Men in white shirts gave her a dollar for the house and said they'd get around to an

Two views of the historic Boston Grain Exchange: from underneath the Central Artery (left); artist's rendering without the artery (below).

The path of demolition to make way for the Central Artery.

appraisal later. The whole neighborhood was destroyed. Not many spoke English well, and nobody knew how to fight it.

When North Brighton had been under concrete for forty years, Salvucci still could get angry about what the highways did to his family. "People didn't matter," he said. "They were going to get pushed out of the way, and that's the way it was in the highway program. 'Insensitive' doesn't begin to describe the brutality of some of what went on."

He heard story after story about all this, and finally he told himself: "This is wrong, and if I'm ever in this field, I'm not going to treat people that way. It's just not the right way to do things."

"BROADER RIBBONS ACROSS THE LAND"

The Salvuccis' sorrow, shared by families wherever massive highways were built through old cities in the 1950s and '60s, can be traced to the years when Americans fell in love with the automobile, the ideal match for a people who perceived absolute freedom of movement as a birthright. In 1900, only eight thousand motorized vehicles were on the roads. Over the following twenty years that figure climbed to 9.2 million. Yet auto owners, yearning to roam, left the city only to bog down in mud and dust. In the entire country, fewer than 150 miles of rural road had any sort of pavement. By the end of World War I, states and counties had applied sand and gravel to lots of roads, but only a very few were paved with concrete or asphalt.

In 1919, General John "Black Jack" Pershing, leader of the Yank expeditionary force that tipped the balance in Europe, came home to a transportation mess. The nation's railroads, 256,000 miles in all, were a shambles. Years of neglect, poor management, and hasty nationalization during the war were making it impossible for the military to move men and equipment around the country. If rails couldn't handle the traffic efficiently, could roads? To find out, Pershing set up a convoy of seventy-nine army vehicles—trucks, artillery tractors, searchlight carriers, ambulances, and motorcycles—and sent it west from Washington, bound for San Francisco. Two young officers volunteered to go along as observers, "partly as a lark and partly to learn." One of them was twenty-nine-year-old Dwight D. Eisenhower.

Eisenhower, a Kansan, knew how primitive western roads were, but even he was struck by the awesome difficulties of moving modern military machinery across the country. Trucks and tractors crashed through old wooden bridges. Wheels sank to their hubs in mud and sand. The convoy jammed up in small towns. By the time the soldiers reached San Francisco they had been on the road for two endless months. The convoy proved what drivers already knew by personal experience—the United States was unprepared for an automotive age.

All through the twenties and thirties, a drumbeat for better highways arose in all walks of life. Sunday drivers complained about clogged two-lane roads in the country. Commuters raged at traffic jams that blocked their way to work. Farmers wanted better access to urban markets. Businessmen called congressmen, insisting that economic growth depended on better movement of goods. Keynesian economists in New Deal agencies said highways would prime the economic pump. President Franklin Roosevelt proposed a system of linked highways that would girdle the country.

In 1939, the notion of a national web of high-speed roads got a boost from the architect Norman Bel Geddes, who bewitched visitors to the New York World's Fair with an exhibit for General Motors called "Magic Motorways," a model of limited-access highways with miniature cars whizzing effortlessly through cloverleaf connections. Thanks in part to Bel Geddes's vision, road-building took on a utopian cachet that survived the wartime freeze on auto production and flourished in the late forties, when the auto factories churned out new models for an eager public and the roadways became more crowded still. In 1944, Congress approved an interstate system in concept but failed to provide sufficient funding.

Thousands of veterans of World War II, including Eisenhower himself, came home from Europe much impressed by the banked, divided highways that Adolf Hitler had built throughout Germany. The autobahns had been tools of war, but many G.I.s thought such roads could serve peaceful aims as well, and it was widely predicted that "superhighways"

Lt. Col. Dwight D. Eisenhower and Maj. Sereno Brett, Wyoming, 1919. On the U.S. Army's first transcontinental motor convoy, in 1919, Eisenhower saw the poor quality of U.S. roads up close.

Road building, Menard, Texas, 1940.

Futurama exhibit, 1939 World's Fair, rendering by Norman Bel Geddes.

soon would speed Americans into a streamlined future of material ease. Advocates made a simple and appealing argument: High-speed highways free of traffic jams would allow anyone with a car to go more places and do more things in less time. The roads were freedom and opportunity expressed in reinforced concrete. The very word someone later coined for these new roads—"freeways"—was a perfect expression of Americans' twin loves of liberty and the open, endless road.

It took a long time for the vision to approach reality. For ten years after World War II, the argument over what to do about the roads raged among politicians and interest groups—builders, automakers, toll-road backers, state highway boosters, farmers, truckers, and the American Automobile Association. The impasse was broken by the soldier who had crossed the country in that jolting convoy of 1919—Dwight D. Eisenhower, elected president in 1952. "The old convoy had started me thinking about good, two-lane highways," Ike wrote later, "but Germany had made me see the wisdom of broader ribbons across the land."

President Eisenhower and the Clay Committee, the Oval Office, 1955. The Clay Committee, made up of influential business leaders and engineers, here delivers its report on how to finance the national interstate highway system.

Eisenhower defined the highway question as a problem of the nation as a whole, not of individual states or regions. Under his leadership, Congress and the interest groups pounded out a compromise that proved wildly popular: A network of divided, limited-access superhighways would be built to standards set by the federal government, but along routes planned by the states. The money would come from a highway trust fund supplied by federal taxes on gas, tires, and other automotive products. And here was the fabulous promise: For every dollar of federal highway money, a state would have to pitch in only ten cents. For governors, mayors, and taxpayers, these "ten-cent dollars" looked awfully good; they could back popular highway projects, yet pay only a fraction of the roads' real cost. And the aid was more or less infinite. Washington was promising to pay whatever it took to get each job done.

In the old cities of the East and Midwest, movers and shakers welcomed the coming superhighways as a man lost in the desert welcomes water. (Eisenhower had envisioned the highways running only *between* cities, not *through* cities, but with so many congressmen representing city constituencies, freeways into, out of, and through cities were inevitable.) Veterans were leaving the cities of their youth to settle in new suburbs spawned by the automobile. Factories, offices, and jobs were following them. Those left behind were poorer and less educated. Business districts and neighborhoods were fading into decrepit old age. Tax bases were shriveling.

In this deteriorating landscape, the only prophets of optimism were the urban planners and architects, touting urban expressways and "urban renewal" as the salvation of the cities. The roads would shrink time and space, the planners promised, benefiting everyone. Poor, old neighborhoods—conveniently stigmatized as "slums"—would be "cleared," their residents dispatched to low-rent high-rises. Mayors and downtown business leaders touted highways as weapons in their competition with the suburbs. Property owners envisioned new development and rising values. The construction unions saw a reliable source of good jobs stretching into the distant future. The

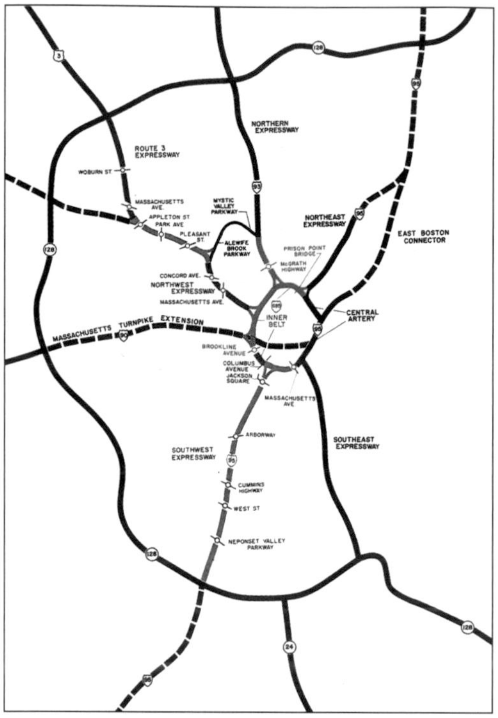

Antihighway activists rebelled against the master highway plan for the Boston metropolitan area that was designed in the late 1940s and 1950s—especially the proposed Inner Belt, which was to encircle downtown Boston. This map shows the master plan as it stood in 1962.

superhighway vision was woven with vivid images of *getting things moving,* and it proved irresistible in city after city.

"WE WERE WRONG"

The superhighway craze rolled into Boston in the fifties with cheers from all sides. With the backing of business and labor, state transportation officials were consolidating projects into a master plan for all of eastern Massachusetts. It was much like plans in many big cities—one or more outer rings or "beltways" with spokes that lanced through city neighborhoods to the hub of downtown. On the planners' maps, all the expressways converged to form a tight ring of concrete called the Inner Belt, just a mile or two from the old city's center. When built, its eight elevated lanes were to cut an arc through the old neighborhoods and towns of South Boston, Roxbury, Brookline, Somerville, Cambridge, and Charlestown. The belt would begin and end at either end of a two-mile elevated highway bisecting the peninsula of downtown Boston. As the downtown structure rose in the mid-fifties, Bostonians realized to their horror it was not just a road but also a wall—an immense, unsightly barrier between downtown Boston and the city's historic North End and waterfront. As a piece of architecture the Central Artery was ghastly. As a road it was little better. Traffic planners eventually would count ten or eleven "rush hours" on the artery daily, even late at night.

Planners believed the Inner Belt would relieve the artery's congestion. Indeed, they saw the belt as the key to the whole system, a detour around downtown yet also the means of rushing suburban commuters in and out of that core.

In the neighborhoods that lay in the proposed path of the Inner Belt, people welcomed the plan as one would welcome a bulldozer aimed at one's front door. But at first, everyone figured the masters of the master plan—the state's Department of Public Works and the powerful business and labor interests supporting it—would have their way, and nobody could do anything about it. Besides, everybody was saying the plan would save Boston. Why fight a losing battle against progress?

People were still coping with the destruction of homes and businesses (known in planners' documents by the genteel euphemism "takings") that had occurred already. The Central Artery alone had claimed a thousand buildings. Many more were sacrificed for an extension of the Massachusetts Turnpike—the project that took the home of Fred Salvucci's grandmother—and a second tunnel from downtown to East Boston. Luxury apartments went up where the heavily Italian West End had been. Scollay Square, decaying but historic, with buildings dating to the pre–Civil War era, was bulldozed to make way for an imposing new Government Center. The neighborhood had been the home of the abolitionist movement. But no one even put up a marker.

Now Bostonians were told new highways would swallow several thousand more homes and businesses, split tradition-bound communities, and fill the horizon with an elevated highway many times longer than the dismal Central Artery. Parks, open space, and wetlands would be blotted out, including a cherished wildlife area south of the city called Fowl Meadow. More old buildings must fall. The automobile would rule Boston.

The Central Artery under construction, 1954. The *Boston Globe* called it the city's "highway in the skies."

In the summer of 1960, a very junior politician from Brookline, Massachusetts, went to Los Angeles to watch the Democratic party nominate another son of Brookline, John F. Kennedy, for the presidency. Michael Dukakis took time to look around Los Angeles, already well along with its own network of urban freeways. Smog burned his eyes. He recoiled at the jammed arteries of cars and trucks. "You looked around at what is really a lovely area," he recalled long afterward, "and wondered, 'What have they done with this place?' I certainly didn't want that for my city."

Here and there around Boston, a few others began to echo Dukakis's complaint. Gradually, one lonely dissident would contact another, then another. As months passed, then years, the tiny network of malcontents grew into a formidable grassroots coalition—the Greater Boston Committee on the Transportation Crisis—plus coconspirators ranging from the Association of Boston Urban Priests to the Black United Front. They were an uncomfortable mixture—waitress-moms, suburban matrons, Charlestown longshoremen, Harvard and MIT professors, East Boston retirees, and scowling radicals, black and white. Yet despite their profound differences, and all the alienation and anger boiling up in Boston and other cities, their anger at the highways drove them together, and

A finished section of the Central Artery in 1954.

Michael Dukakis campaigning, 1966.

Antihighway protest, Boston, 1964.

they waged guerrilla war on the master plan all through the sixties. They delayed a permit here, called for more study there, filed a lawsuit somewhere else. They didn't stop every project. But as the years passed, most of the planned Inner Belt and the local section of I-95, known in Boston as the Southwest Expressway, remained unbuilt.

In the state legislature, one of their key allies was Michael Dukakis. But among the lawmakers, as elsewhere in town, it was heavy lifting indeed for Dukakis to argue that the vaunted master plan was a route to sorrow and that an abstraction called mass transit offered Boston a better future than King Car. Dukakis was smart, but he was a lawyer, not an engineer, and he kept looking for somebody who could talk back to the builders and planners in their own arcane tongue. Then, as Dukakis remembered it, somebody told him: "There's this young guy who's an MIT-trained engineer and he's working at the Boston Redevelopment Authority as a transportation planner. He's just come back from a Fulbright in Italy. His name's Salvucci. Why don't you get ahold of him? I think he agrees with you."

Dukakis's source had it right. Fred Salvucci was living with his wife and children on the third floor of a Brighton "three-decker"; his parents lived on the first floor, his sister and her family on the second. By day he was drawing urban redevelopment plans. By night he was volunteering as a technical consultant to the anti-master-plan agitators. "He was peculiarly suited to this function," a journalist wrote a few years later, "for he embodies much of what Boston is. He is an ethnic and an intellect. He speaks English and Italian, both often in a mumble. His English is really Italo-American; the 't's' are spit out, and the consonants are hard and decisive. . . . This same man is a graduate of MIT and a former Fulbright scholar. He is regarded as a valuable transportation technician. A public speaker, he is not; a convincing exponent . . . of the social costs of progress, he is."

When Kevin White, Massachusetts's secretary of state, became mayor in 1967, he picked Salvucci to manage the first of his "Little City Halls," this one in Italian East Boston, where Salvucci helped the neighbors fight the encroachments of Logan Airport.

Powerful figures began to lend their support to the antihighway movement—at Harvard, Professors Daniel Patrick Moynihan and John Kenneth Galbraith; in Washington, Senator Edward Kennedy and Congressman Tip O'Neill of Cambridge, a rising leader among

House Democrats; at City Hall, the mayor himself, eager to run for governor and seeing transportation as an emerging hot-button issue, and his chief of staff, Barney Frank, a shrewd young liberal whose antipathy toward the highways sprang from concern for shattered neighborhoods, not urban esthetics. "If the whole world were like Bayonne, New Jersey, it wouldn't bother me," Frank said later. "It wasn't the sanctity of Fowl Meadow but the effect on the poor that bothered me about I-95. I'm not a druid. I don't think trees have an independent value."

Massachusetts governor John Volpe and California governor Ronald Reagan, 1968. Both would play a role in the Big Dig story.

Mayor White's support, though a long time coming, proved critical. The son of a Boston Irish pol and a mother who read Willa Cather and Tolstoy, White moved easily from blue-collar parish picnics to Harvard sherry hours, and he knew the persuasive arts well enough to cool down an angry black crowd at a James Brown concert in the hours after Martin Luther King, Jr., was murdered. As the highway fight escalated in 1969, White escorted more than one caller to a window overlooking the Central Artery and said: "Look at that thing. It's a goddamn Chinese wall. I'll be damned if I let one of those ruin my city while I'm mayor."

Yet at just that moment, events were shifting in favor of the master planners. Richard Nixon, just inaugurated as president, tapped Massachusetts governor John Volpe to become U.S. secretary of transportation. Volpe was a highway builder; he had directed the state's Bureau of Public Works and overseen much of the highway master plan. In fact, he had signed the papers that put contractors to work on a section of the Central Artery. And Volpe's successor as governor was his fellow moderate Republican, Lieutenant Governor Frank Sargent, *another* ex-head of the Bureau of Public Works. Volpe went off to Washington, committed as deeply as ever to the Inner Belt. But Frank Sargent, as Boston soon learned, was not your typical highway builder.

As Mayor White prepared to run against Sargent for governor in 1970, Barney Frank, Fred Salvucci, and other mayoral aides urged White to seize the banner of the antihighway forces. White made his move, demanding that Sargent announce "an immediate halt to any land-taking, demolition, or construction now taking place or contemplated for new highways," which White blamed for destroyed homes, wrecked neighborhoods, polluted air, disappearing open space, and the rising public view "that government is an unfeeling, unthinking monolith." An editorialist at the *Boston Globe* suggested the mayor's proposal was "about as practical as a petition to the Weather Bureau for a moratorium on snow in New England."

But as Bob Dylan said in that era, you didn't need a weatherman to know where the wind was blowing, and Governor Sargent could read political skies as well as anyone. As an old-family Bostonian, an MIT-trained architect, and an ardent conservationist, he had his own doubts about urban freeways, despite his tenure as the state's chief road-builder. When protesters crowded into the governor's office to shout about the Inner Belt and the Southwest Expressway, they found Sargent actually listening to them, apparently devoid of the officeholder's congenital urge to escort unhappy constituents out the door as quickly as possible.

Soon Sargent set up a task force under Alan Altshuler, a political scientist at MIT and an authority on transportation policy who was beholden to neither the pro- nor the anti-highway forces. For many months Altshuler's task force listened to protesters from one side of the firing line, engineers and planners from the other. Anti-highway advocates argued that the Inner Belt, far from reducing traffic in the central city, actually would bring a dramatic increase, forcing yet another round of demolition to make way for more parking spaces. Their arguments persuaded even prohighway businessmen on the panel that the Inner Belt made little sense. Then, in startlingly blunt terms, the task force told Sargent the highway planning process in eastern Massachusetts was "pathological," "a highway juggernaut," "a great mindless system charging ahead," fueled not by public need or even solid data but by the simple lure of ten-cent federal dollars. The task force urged Sargent to halt the key projects, including the Inner Belt, pending a broader study of the region's true transportation needs.

For a month Sargent thought about it. He was under enormous pressure. Kevin White was making the issue his own, and the anti-highway groups were winning sympathy not just in the Democratic city but in the Republican suburbs. Yet many in the business community, Sargent's natural constituency, had put big money on the line along the routes of the planned highways. Thousands of construction jobs depended on the decision. And state planners were telling Sargent that to pause now, with only part of an integrated highway system complete, would make traffic worse than ever.

Finally Sargent asked for statewide television time. Facing the camera on February 11, 1970, he declared: "I have decided to reverse the transportation policy of the Commonwealth of Massachusetts. . . . Nearly everyone was sure highways were the only answer to transportation problems for years to come. But we were wrong."

In Washington, John Volpe went through the roof of the U.S. Department of Transportation. But in the fall election, Sargent defeated White and his running mate, Michael Dukakis, and named Alan Altshuler as the state's first secretary of transportation. Under Altshuler, an entirely new blueprint emerged. It buried the Inner Belt for good and called for a major commitment to mass transit. Sargent agreed to it.

"It was an incredibly courageous thing for Frank Sargent to do," Salvucci said much later. "I'm a Democrat. I don't say many good things about Republicans. But he was a great man."

Sargent said two highway ideas still deserved a close look. One was to build a third tunnel under the Inner Harbor from downtown Boston to Logan Airport. The other was to figure out a way to fix the Central Artery by placing it underground. Those ideas heralded the end of Fred Salvucci's career as an advocate and the beginning of his career as a builder. Without anyone knowing it, the victory of everyday people over the interstate highway program had planted the seeds of the biggest urban highway project in U.S. history. And Salvucci, the antihighway guru, would lead it.

Governor Frank Sargent swears in Alan Altshuler as Massachusetts's first secretary of transportation, 1971.

"A WONDERFUL VISION . . . "

The insurgents celebrated. But Boston remained mired in automotive quicksand. Sargent's proposal to build a tunnel to Logan Airport stalled. Michael Dukakis, elected governor in 1974, continued the overhaul and expansion of the city's subways that Sargent had begun. But at street level, Boston's perpetual rush hour persisted year after year, through the reigns of Governors Sargent (1969–75); Dukakis (1975–79); Edward King (1979–83); and Dukakis again (1983–91). Meanwhile, the Central Artery was deteriorating under the pounding of coastal weather and Boston traffic; soon it would be dangerous. It had to be replaced. But the contesting parties—Democrat and Republican, prohighway and antihighway—remained deadlocked. Their decades-long battle is described in detail by David Luberoff and Alan Altshuler in *Mega-Project: A Political History of Boston's Multibillion Dollar Artery/Tunnel Project.* It will doubtless be remembered as one of the great struggles in American urban history.

Through all the battles, Fred Salvucci could be seen moving

Rumney Marsh, 2000. As part of the Big Dig's massive efforts at environmental mitigation, the state paid for the restoration of an eighteen-acre section of the marsh, which had been a wetland before the highway development of the 1960s. It was partially destroyed during preparations for a never completed highway. As compensation for filling in portions of Boston Harbor and capping the former Boston landfill at Spectacle Island, the state restored this important habitat.

irony: the old antihighway crusaders who killed the Inner Belt were asking Mr. Highway to back their own highway plan. But Volpe listened closely, and when his petitioners had finished, he told them he had reviewed plans for the Central Artery in the fifties and said, "This is a giant mistake." It had been too late to change it, Volpe told Salvucci, but he had regretted it always, and he would do what he could to help.

But federal highway administrators fought the project at every step. They objected on legal and technical grounds, and they doubted the project would solve Boston's traffic problem. They delayed approval of Massachusetts's environmental impact statement. They said the project failed to meet the requirements for interstate funding. And they got the whole matter referred to Congress. Salvucci attributed the delay solely to politics. Because the state had rebelled against highways in favor of mass transit, he said, key highway officials in Washington harbored "a deep-seated dislike of Massachusetts in general and of Frank Sargent, Alan Altshuler, Michael Dukakis, and Fred Salvucci in particular." Whatever their motives, the administrators wrapped the Boston project in red tape for four years.

As the parties wrangled, Speaker O'Neill responded with red tape of his own, locking up monies for many other transportation projects around the country; if Boston didn't get its funding, no one would. Salvucci and his aides brought in top Republican lawyers to make their claims; lobbied officials in other states to support their case; and debated endlessly with Federal Highway Administration staffers over cost-benefit ratios and travel-time savings.

At last a compromise was reached. The state would depress the central segment of the artery with its own money and other federal funds. Interstate dollars would pay for most of a new Charles River crossing at the artery's north end; an interchange and tunnel at the south end; a seaport access road in South Boston; and the new airport tunnel.

But President Reagan was having none of it. By now it was 1987, and Reagan was under seige. On one side he faced the Iran-Contra scandal. On the other, critics were saying he was out of touch, maybe even senile. Democrats had regained control of the Senate, and in the wake of Tip O'Neill's retirement, the new Democratic speaker, Jim Wright of Texas, was declaring his party would now control domestic policy. In this setting, Reagan was in no mood to hand his enemies a giant plum in the hometown of Tip O'Neill. When the House and Senate approved the broad transportation bill that included the artery/tunnel project, the president vetoed it. Reagan singled out the Boston project for scorn, saying, "I haven't seen this much lard since I handed out blue ribbons at the Iowa State Fair."

Boston's forces had to override the veto, or the entire project would die. Yet for Reagan the matter had become much more than a skirmish over roads. It was a last stand against pork-barrel politics and a test of his wavering power. He told Republicans that if they couldn't sustain his veto this time, Democrats would trash seven years of Republican achievements. The House voted to override by a wide margin, and attention shifted to the Senate, where seventeen Republicans had voted for the highway bill originally. Thirteen said they would stick with it now, either because the bill would raise rural speed limits to 65 mph—a highly popular measure, especially in the West—or because their states would gain their own funding for highways or mass transit under the bill. Barney Frank, now a U.S. congressman with twenty-five years of service in Boston's highway wars, moved to lock up their support. He let the Republicans know that urban Democrats in the House who had backed the higher speed limit would not do so again if Republican support for the highway bill wavered in the Senate. A Salvucci aide later called this "a delightful piece of reverse blackmail. That is, 'You have something now, but if you don't do what I want, you won't have it later.'"

Only one Democratic senator had voted against the bill originally—Terry Sanford of North Carolina, a former governor and president of Duke University who had just taken his Senate seat that year. He had opposed the bill because his state received less money in highway funds than it sent to Washington in gasoline taxes—an inequity that angered many—and he had promised to support a presidential veto. Now the bill to override the veto came down to one vote—his.

First Sanford voted "present," favoring Boston. Then he switched his vote to "no," favoring Reagan. The president's backers erupted in celebration. But in a quick parliamentary maneuver,

Senator Terry Sanford, 1990. His vote was crucial to the Big Dig.

Robert Byrd, the Democratic leader, was able to call for another vote on the measure. That gave the Democrats a little more time to work on Sanford.

Seldom has the art of political logrolling been so conspicuous. In the House, Democratic leaders went to their friends in the North Carolina delegation and began to talk about tobacco, the tarheel state's principal crop. They had voted for federal subsidies to tobacco farmers through the years, they said, but those subsidies were becoming an embarrassment. And if Terry Sanford could not be persuaded to support this very worthwhile highway bill, then the tobacco subsidies would have to die, too, and Democrats would become heroes in the war on cancer. Very soon Senator Sanford concluded that tobacco meant more to his state than a couple of roads, and he changed his vote. North Carolina would keep its tobacco subsidies, and Boston would have its interstate funding.

On most Tuesday afternoons in 1988–89, Salvucci sat down with twenty or so engineers and planners to work on the designs for the difficult crossing of the Charles River. They stayed until ten or eleven or midnight, eating pizza and sketching until they resolved the issue at hand or gave up for the night. As time went on, a letter of the alphabet was given to each successive plan, or "scheme." They drew thirty schemes in all. The only one they thought could work was the twenty-sixth—thus, the sinister-sounding "Scheme Z." That was the one they took to the public in 1989. "We never should have used the word 'scheme,'" said Bob Albee, "and we never should have settled on 'Z.'"

But it wasn't the name that turned Scheme Z into a planner's nightmare. It was the structure itself. On the Charlestown side, a north-south highway, I-93, meets a series of roads that form an east-west highway. If the Big Dig were being built in midwestern farmland, engineers would build a cloverleaf interchange to connect the two roads. But in downtown Boston there isn't room for that. So in Scheme Z, the engineers took the loops of the cloverleaf and stacked them on top of each other. The result was a sixteen-lane structure seven hundred feet wide and eleven stories high, plus three bridges to accommodate all the ramps and traffic. The interchange would cast a shadow for a quarter of a mile. An Environmental Protection Agency official said it would be "the single ugliest structure in New England."

When a model of Scheme Z was unveiled in the summer of 1989, Charlestown and next-door Cambridge revolted. They blasted Salvucci for sticking this aboveground monstrosity in working-class Charlestown while he planned to beautify downtown Boston by burying *its* monstrosity underground. The city councils of Boston and Cambridge came out against Scheme Z. Environmentalists fought against it in the permitting process. When Salvucci promised $75 million for parkland along the Charles—another attempt at "mitigation"—opponents said he was "putting a mustache on Frankenstein." Many opponents called for the crossing and interchange to be built underground, like the Central Artery, but Salvucci's people said that would raise the project's cost by $1 billion, maybe more. As the Scheme Z storm intensified, other opponents arose to attack the entire Central Artery/Tunnel Project. In a *Boston Globe* essay, a critic called it "a mistake of unprecedented magnitude . . . a highway out of the 1950s . . . catastrophic from an environmental perspective." Some critics were environmentalists and concerned citizens in the areas most affected by Scheme Z. Others were lawyers and technical experts funded by the owners of a long-term parking lot next to Logan Airport. The new tunnel would take a fifty-foot slice of the lot. Until they reached a settlement with the state, the owners led the revolt against the Big Dig.

A supporter of Salvucci responded: "The critics are like someone who is watching a guy on a high wire standing on one foot, balancing a grand piano . . . in the air, playing a Beethoven sonata on the piano while sipping soda from a straw and hopping up and down and when he finally gets to the other side, saying, 'He's not a very good piano player, is he?'" Opposition to Scheme Z reached its peak just as a crucial deadline was looming. After his defeat in the 1988 presidential election, Michael Dukakis had chosen not to seek re-election as governor in 1990. Both candidates to succeed him—Boston University president John Silber, the Democrat, and William Weld, a former U.S. attorney and a moderate Republican—had endorsed the Central Artery/Tunnel Project. But Weld, who won, was approaching the matter cautiously, especially the artery. If Scheme Z didn't receive the official blessing of John DeVillars, the secretary of environmental affairs, by January 3, 1991, when Weld took office, the project might lose so much momentum that it would stall for good. For the first time, a major attack was leveled at Salvucci himself. Environmentalists inside and outside government charged that

Presidential preference poll of New Hampshire voters, Page 3

The Boston Globe

FRIDAY, APRIL 3, 1987

104 Pages • 25 cents

Senate overrides veto of road bill

Artery, harbor project due to span 11 years

President still a force to be reckoned with

Reagan loses by 1 vote; speed limit is eased

The Boston Herald

WEATHER

TODAY'S TV: Page 37

25 Cents

Friday April 3, 1987

RELIEF!

The front pages of the *Boston Globe* and the *Boston Herald*, Friday, April 3, 1987.

Salvucci loyalists were pressuring the Department of Environmental Affairs to rush the approval through, even to change parts of its report to favor Scheme Z—an accusation Salvucci denied. "If we don't finalize this thing and get into construction," Salvucci warned, "we can lose this project." For him alone, that would mean twenty years of work down the drain.

On January 2, 1991, the last day of the Dukakis administration—and thus the last day of Salvucci's stewardship of the project—John DeVillars called a news conference. He made a major concession to the critics by strongly recommending a broad new review of Scheme Z. But he approved the necessary environmental permits, and he refused to speak a word against the project's principal author. His throat suddenly tightening, DeVillars declared: "No one should lose sight of Fred Salvucci's incredible contributions to serving the public good."

Three years passed as the Weld administration reconsidered the Charles River crossing. Hopes focused on all-tunnel designs, but these raised serious questions of river pollution, cost, and traffic safety. So the planners turned to bridge-and-tunnel ideas and bridge-only ideas, hoping to find a scheme superior to Z—but it was clear that any bridge-only design would entail one of the widest traffic structures in the world, a feature that environmentalists and architecture critics had blasted during the Scheme Z fight.

Engineers have a saying: "If you can't fix it, feature it." If they had to have a very wide bridge, planners decided it also had to be very beautiful. So they went to Christian Menn. Like Othmar Ammann, Menn was Swiss. Born in 1927, he had established a reputation as perhaps the premier bridge designer in the world, departing from the long-span suspension bridge with striking designs in concrete, mostly in Europe. Salvucci had consulted him earlier. Now, for the Charles crossing, Menn drew a lovely cable-stayed bridge comprising two inverted-Y towers and skeins of diagonal cables. The form was not only strikingly beautiful but unique in the world; it was the first great asymmetrical bridge, with two extra lanes cantilevered to one side, making ten lanes in all. When Menn, at a presentation, lifted a model of his design out of a plywood box, one of those present recalled, the reaction was a universal "Wow." With Menn's design, the Charles River crossing went from being the project's leading nightmare to its signature and symbol—the one main element that rises above the ground, providing the landscape of

Boston with a work of surpassing beauty even as it performs the workaday task of moving people across the river.

EXECUTION

Engineers had built or were building bigger projects—the Suez Canal, the Panama Canal, the English Channel Tunnel—but never such a big project through the middle of a city. "Engineers can *design* anything," said Mike Lewis, a key engineer on the project. "But when somebody says, 'All right, now go and do it,' that's when the difficulty starts. How do we actually *build* a tunnel under an elevated highway and keep traffic running at all times? And, by the way, we have to keep the pedestrians moving at all times. And we have to keep the shop owners and the businesses in the same place. And we have to keep people sleeping at night that live right next to it. It's those kinds of complexities that induce panic for engineers."

To transform the evolving design into reality, Salvucci, before he left office, had hired a consortium of engineers and construction managers from two large firms with experience around the world in massive projects: the Bechtel Corporation (one of Hoover Dam's Six

Model of Scheme Z. Vehicles would have to cross the Charles River twice to reach certain destinations.

Model of Charles River Bridge, 1992.

Charles River Bridge construction, 2000. Menn's design of the inverted Y-shaped towers is intended to reflect the Bunker Hill Monument in nearby Charlestown.

cylinders that would become the roadways in the tunnel. Then each section, big as a ship, was loaded onto a barge and towed up the coast to Black Falcon Pier in South Boston. There, workers lined the tube sections with steel-reinforced concrete, then sealed the ends to make them airtight.

Under the harbor's surface, other workers had been making a resting place for these leviathans. Using the biggest dredge in the world, they dug a trench across the bottom of the harbor. The job yielded nearly a million cubic yards of dirt, which was towed to nearby Spectacle Island and dumped on top of a landfill; this would be the ground under a new city park. When the trench was complete, first one tube section was gently lowered into it, then another. The two were coupled together; then another section came down, and so on until all twelve sections were in place, stretching across the harbor. Then the trench was covered over. Much work remained be-

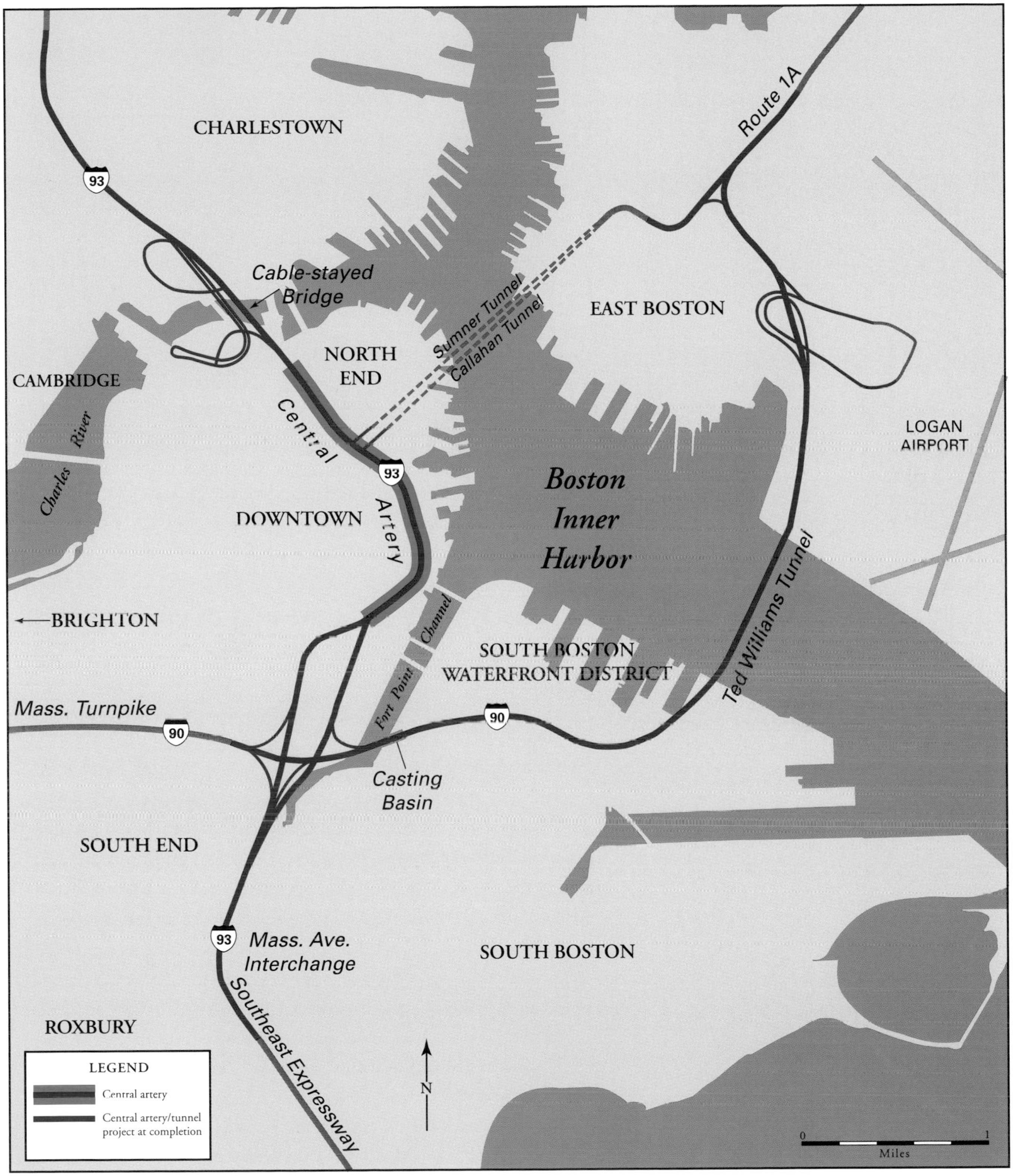

The Central Artery/Tunnel Project in its final configuration.

In a 1993 "float-out," a section of the Ted Williams Tunnel is prepared for submersion.

fore a car could drive through it, but it was a tunnel—a steel, immersed-tube tunnel under Boston Harbor.

The tunnel—named for the Red Sox Hall of Fame outfielder Ted Williams—opened in December 1995. For its first six years in operation it would carry only buses, taxis, trucks, and airport limousines. General traffic was to begin once the tunnel was connected to Interstate 90 in 2001.

The Central Artery

Depressing the artery presented an altogether different and even more difficult problem in tunneling. Here the engineers had to dig a tunnel under the old elevated roadway, and the new, underground roadway had to be wider than the old roadway above. The job would have been vastly easier if they could have torn down the artery first, then dug the tunnel. But the artery had to stay put to carry traffic.

The engineers said it was like performing open-heart surgery while the patient was playing tennis.

The artery stood on a foundation of eighty-six underground columns that rested in turn on piles—stout posts that ran down to bedrock. These old columns and piles stood right in the path of the planned tunnel. Somehow the engineers had to get the columns and piles out of the way without the artery above collapsing.

They solved the problem in stages. First, they excavated a trench three feet wide and up to 120 feet deep, down to the bedrock under the city. The trench was dug in sections called panels, each of them ten feet long. As an excavating machine dug dirt out of the hole, workers pumped a polymer-and-water mixture called slurry *into* the hole. The gooey slurry was thick enough to keep the walls of the hole from caving in, but thin enough for the excavating machine to plunge through and keep digging. Every five feet along the trench, they stuck a massive steel column called a soldier pile straight down into the slurry. Next, they pumped concrete down into the bottom of the trench between the soldier piles. The concrete, filling the trench from the bottom, pushed the slurry up toward ground level, where it was skimmed off for reuse down the line. It took two days to build one ten-foot-long panel. When concrete filled the trench to a certain point and hardened, the engineers had a ten-foot-long steel-reinforced wall in the ground.

Slurry wall construction. A clamshell excavator digs a trench, 3 feet by 10 feet and as much as 120 feet deep, which is then filled with the liquid slurry.

The engineers built two of these buried walls along the entire length of the Central Artery. Then they laid great beams from one wall to the other. Beams and slurry walls together made a new foundation for the Central Artery. The engineers transferred the roadway's load from the old foundation to the new. Then they dug and blasted and hauled out the old foundation. Finally they could dig their tunnel within the slurry walls.

The workers dug a massive trench, built the tunnel in the trench, then covered it over. Meanwhile, the slurry walls were not only holding up the old Central Artery but also keeping the water of Boston Harbor from seeping into the tunnel. All this was done with a minimum of disruption to drivers and pedestrians overhead. Wherever the builders got below the surface, they installed a concrete deck over their heads, at ground level. People passing on the street soon forgot a tunnel was being excavated under their feet.

The Fort Point Channel

The most difficult engineering challenge of all emerged early, when Salvucci was still in office. It turned up in the Fort Point Channel, the narrow arm of water between downtown and the South Boston waterfront district, where Bill Reynolds had gotten his idea about the tunnel route to the airport. Five hundred feet across, the channel is the remnant of a much larger body of water. Over two centuries Bostonians filled most of it to make more space, mostly for wharves, rails, and factories. The plan called for I-90 and I-93 to come together in a massive interchange at the channel's western edge. From the interchange, I-90 would proceed eastward through a tunnel under the Fort Point Channel, then bend around under South Boston to join the Ted Williams Tunnel a mile or so to the east. To complicate matters, the Red Line subway, an aging structure built in the early 1900s, ran under the channel, too.

Powerful neighbors stood on the channel's banks—the U.S. Postal Service, handling every piece of mail going in or out of Boston, and the Gillette Company, maker of razors, one of the city's biggest employers. Both worried about the havoc that underground vibrations might cause in their sensitive machinery. Those were big problems in mitigation. But the key problem was whether the tunnel under the channel could be built at all.

One day a delegation of engineers came to Salvucci and said: "This section past the Fort Point Channel and the Gillette Company—we don't know how to build it."

"What do you mean you don't know how to build it?" Salvucci said. "We've been through dozens and dozens of hearings. We've done all this environmental analysis. We've got approvals. Now you're telling me there's a problem building it? You've *got* to build it."

Normally, they would create a tunnel like this using the same immersed-tube technique that created the Ted Williams Tunnel. But that would mean floating barges up the Fort Point Channel. They couldn't do that. There were three low bridges in the way—the only connections between downtown and the sprawling South Boston piers. Nor could they build a steel mill in South Boston just for the purpose of making this tunnel. Salvucci listened for a while and realized the engineers had come up against a problem that might just be insoluble.

"You guys are the best in the world," he told his visitors. "You've got to come in with a solution. Don't come in with a problem."

Work on the new Central Artery proceeds below street level as city life goes on overhead.

So they brought in a solution named Lou Silano, a leading engineer at Parsons Brinckerhoff. Silano reminded Salvucci of his own father and other men he had looked up to as a kid—about five foot nine, chunky, no bullshit. His speech was pure Brooklyn. Before joining the Big Dig, Silano spent twenty years building long-span suspension bridges. He managed construction of the long tunnel under the James River at Hampton Roads, Virginia, and other such projects around the globe. Yet even Silano had been awed when he came to Boston at the beginning of the Big Dig's design process. He sat in a room with two colleagues and sketched a schematic diagram of the steps required to build the entire project. The diagram soon covered all four walls. "You felt all alone—three people looking at a job that was humongous," he said later. "I've worked on many big, big projects. But this was enormous." But Silano plunged in and soon became one of the most respected figures in the project. "When Lou spoke," Bob Albee said, "you listened. He was not a polished consultant. The way he wanted it done, it got done. But he was always right, and you always felt good when you were dealing with Lou Silano."

Silano studied the Fort Point site and somehow saw it in a new

Construction of casting basin in the Fort Point Channel, circa 1999. Workers had to excavate more than 450,000 cubic feet of dirt to construct the basin.

way. If he couldn't use steel, how about concrete? Concrete tunnel tubes never had been used in the United States, but they had worked elsewhere. And if they couldn't be brought up the channel by barge, why not build them in a casting basin right on the spot? And since the tubes had to be floated into position down under the channel, why not build the casting basin in a great pit below ground level, next to the channel? When the tubes were finished, the pit could be connected to the channel, and the tubes could be floated into place and connected to make the tunnel.

Though the details were infinitely more complex, that, in essence, was Silano's solution. A giant casting basin was dug at the edge of South Boston. It was 1,000 feet long, 300 feet wide, and 60 feet deep, big enough to hold three *Titanics*. By 2000, the six giant rectangular sections of the tunnel were cast. Steel bulkheads were installed across the ends of each section, making it just buoyant enough that when the basin was filled with water, the section could be floated out into the channel. Once in position, crews filled ballast tanks with water, pushing the section downward toward the trench. Then the sections were painstakingly maneuvered together and sealed.

When the engineers discovered that the soil under the Fort Point Channel was dangerously soft, they made new soil. Crews would inject water into the soil to liquefy it. Then they pumped cement grout into the mixture. With long, drill-like augers, they mixed the mud and concrete together to make a new substance called "soil-crete," which would be stable enough to withstand the pressures of excavation. Engineers also had to protect the eighty-year-old Red Line subway, which in some spots would run just six feet under the new highway tunnel. Any pressure on the subway tunnel could result in catastrophic damage. So crews drilled shafts for 110 concrete piles deep into the bedrock; the highway tunnel would rest on these, leaving the Red Line untouched.

"A SYMBOL OF TECHNOLOGICAL POWER"

In 2001, the peak year of work, the builders of the Big Dig were completing $120 million worth of construction a month. When it was all done, they would have excavated 14 million cubic yards of earth from downtown Boston, enough dirt to fill a professional sports stadium fourteen times. But Bostonians had long since grown tired of the fabulous statistics.

The casting basin in the Fort Point Channel was big enough to hold an aircraft carrier.

The date when it would all be done stretched into the future—2005 was the latest estimate. And the costs rose to staggering dimensions. In 1983, state officials estimated the cost in current dollars, without mitigation or right-of-way costs, at roughly $2.5 billion. By 1990, taking projected inflation, mitigation and right-of-way costs into account, estimates had risen to more than double the original amount. By 2001 the total cost was projected to reach roughly $14 billion. Analysts attributed the enormous increase to various factors—design changes, mitigation costs, the inflation that occurred through the long years of conflict over the plan. Critics were outraged at the escalation; supporters said it was unavoidable, pointing to similar increases for the Alaska oil pipeline, Los Angeles's subway, the new Denver airport, and the English Channel Tunnel. Certainly the price will give pause to any city considering a similar project. Alan Altshuler called the Big Dig "a wonderful asset for the Boston region." But, he said, "it would never have been built if people had realized how much it was going to cost."

Salvucci, who watched the construction go forward from a faculty post at MIT, said that in spite of the great cost, large cities have little choice but to fix their problems as Boston has. In the summer of 2000, the historian Thomas Hughes begged Boston officials not to trim the finishing touches that could make the city proud in years to come. "Managerial lapses, engineering reverses, and cost overruns are inevitable in such projects," he said. "The possibility is to minimize, not eliminate them." The Big Dig, Hughes said, "will eventually become a symbol of the country's technological prowess."

Fred Salvucci planned for a future that others didn't quite see. For example, he was insistent about putting an exit ramp in a desolate railroad yard in the middle of down-on-its-luck South Boston. Many people, especially in Washington, asked why a railroad yard needed a freeway ramp. Salvucci believed it could transform the area's economy, even affect the future of the city as a whole. And it appears he was right. The interchange spawned a conference center, major hotels, a courthouse, and a large pier development—a phoenix from the ashes that many now see as an extension of downtown Boston. "Because of the investment," said Bruce Katz, an expert on urban planning at the Brookings Institution, "Boston is positioned to be the jewel of American cities in the next century. It gets to the notion that maybe we don't have to look at dying cities as inevitable."

A Manmade City

For all its immensity, the Big Dig shrinks in comparison to the construction of Boston itself, a project that has gone on for nearly four hundred years. Human hands have made the landmass of the city four times larger than the original, claiming acreage yard by yard from the sea—a process superbly documented by a group of scholars in *Mapping Boston* (1999).

Puritans built the city's earliest incarnation on the three hills of the Shawmut peninsula in Boston Harbor. This delicate landmass would have been an island but for a skinny neck connecting it to the mainland. Other peninsulas and islands surrounded Shawmut like random inkblots. These landforms were flanked by broad mudflats, or tidal basins, that rose into view when the tide went out and disappeared when the tide returned. By Massachusetts law, a landowner's property lines extended across the flats all the way to the low-tide mark. If you could build up that soggy acreage so the tide couldn't cover it, you would increase your usable land. That incentive launched the Boston tradition of land-making.

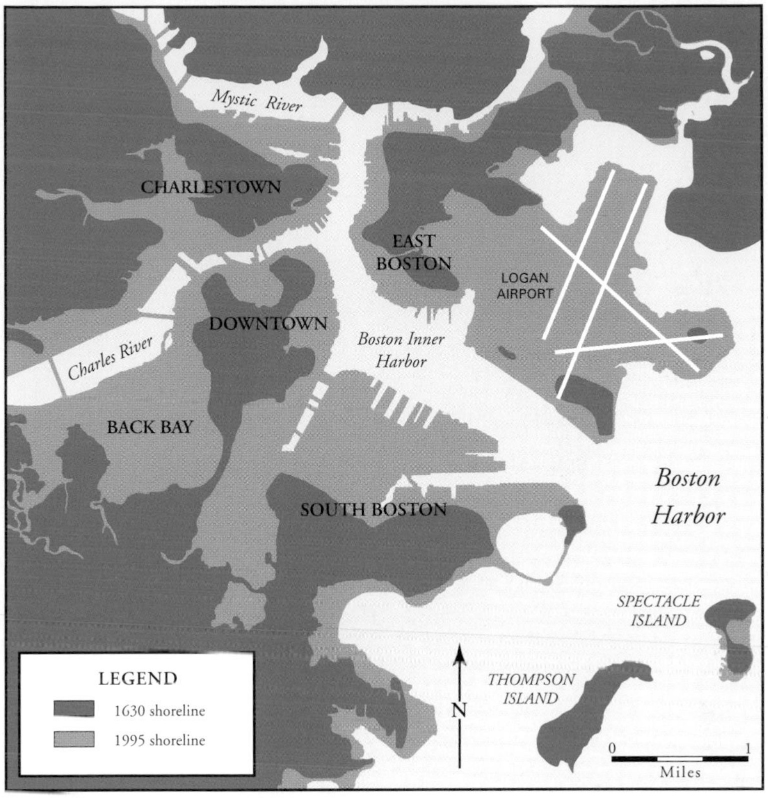

It began in the 1600s and 1700s with a bit-by-bit process called "wharfing out." In this colonial town where seafaring was all, wharves were built out from the land. Then, as time passed, the slips between the wharves would become too small for newer, bigger ships. So the slips would be filled in with earth, and new wharves would be built. Boston's famous Faneuil Hall, for example, sits atop earth that was dumped onto the old town dock about 1730. Elsewhere in the city, people created land by laying down a grid of timbers on a section of mudflat, then dumping earth over the grid. As the work proceeded, Boston's original hills diminished; workers with picks and shovels were mining them for dirt and gravel, then hauling the earth to landfill sites in horse-drawn carts.

The land-making tradition expanded for various reasons, both social and commercial. When, in the 1840s, the great influx of Irish immigrants caused the city's population to grow by nearly 50 percent, city leaders worried that well-to-do Yankees would flee beyond Boston's borders. So they created new residential blocks by dumping fresh earth and gravel along the edge of South Bay, a development that became known as the South End.

An ingenious method of creating power led, in time, to the filling-in of the broad inlet called Back Bay. From 1818 to 1821, a dam was built across the mouth of the bay; then a wall called a "crossdam" was built across the bay, dividing it into two basins. The high tide would flow into the upper basin. Then, as the tide withdrew, water would run down into the lower basin through sluiceways in the crossdam, creating power for mills at the site. But in time the dams also trapped intolerable levels of sewage. "The surface of the water . . . bubbles like a cauldron with the noxious gases that are exploding from the corrupting mass below," one fascinated observer noted. So the filling-in process began here, too. Filling Back Bay required some twenty years of labor, with gravel brought in from Needham, nine miles distant, on a rail line specially constructed for the project. For a time three trains made twenty-five trips per day. When it was done, wealthy Bostonians moved into a space where only coastal sea creatures had lived for eons.

The land-making process continued throughout the nineteenth and twentieth centuries. To accommodate more immigrants arriving in the 1880s, the East Boston Company went to work on adding acreage to that peninsula—a project that eventually became the largest of all, big enough to accommodate the modern Logan Airport. Concerns about maintaining Boston's position as a leading port led to massive land-making in South Boston. Railroads and railroad stations demanded yet more space. In the 1970s and 1980s, the University of Massachusetts added to a piece of land called Columbia Point—itself largely a manmade landmass—and built its new Boston campus there. At the tip of the point the John F. Kennedy Presidential Library was constructed.

Thus, a landmass that once resembled a delicate hand with splayed fingers looked from the air by 2000 like a muscle-bound forearm and fist, with even bulkier manmade landmasses surrounding it.

But other cities are wondering if any improvement in their own transit systems could possibly be worth so much time, trouble, and money. The Big Dig's planners have urged them to step inside the project, to learn what was intended and what is being accomplished, before they shy away. David Luberoff, a Harvard researcher and coauthor of *Mega-Project,* cited positive and negative lessons alike. "There are parts of the Big Dig that will be a model for other cities and parts that will be a cautionary tale," he said. "It's a model in the sense that its planning process was very open. It's a model in the sense that extraordinary efforts were made to see if you could provide new infrastructure in a way that didn't seriously harm other people. It's a cautionary tale about how that kind of approach can get out of control."

Whatever Boston's judgment on whether the Big Dig was worth the money and the wait, there is no question that history will judge it an engineering feat for the ages. Probably no other single project has incorporated so many difficult techniques and designs—"every imaginable type of construction that's going on in the world," Lou Silano said. And certainly no project of such magnitude has ever tried so hard not to bother its neighbors, however real the inconveniences and however fed up some neighbors became.

The engineers themselves remain in awe of what they are doing. "In my book, the Panama Canal was an easy job," Silano reflected. "It was just done at a time when we didn't have the capabilities we have today. It's really a dredging job, a digging job. I think the Central Artery was a much more challenging project." Students of urban politics are equally awed. "The technical work of this project is phenomenal and extraordinarily complicated and amazing," David Luberoff said. "But in many respects the political work that preceded it is even more amazing."

Forty years have passed since Fred Salvucci said there ought to be a better way to build highways in a big city. "I think fifteen years from now the project is going to be largely forgotten," he said as the Big Dig neared completion. "Most of it's underground. People will see the bridge over the Charles and say, 'Oh, that's a unique bridge,' and wonder how it came to be. But nobody ever talks about remarkable engineering achievements like the New York subway system. I mean, do people go into the subway every day and say, 'Wow, what a great subway system! Imagine the work! It's an incredible engineering achievement!' You know, people get up in the morning, they get their cup of coffee, and they go ride the subway. And you hope it works. The Big Dig is going to be a lot like that. I'm really proud to have been a part of it. I think there are thousands of other people who are going to feel that kind of pride. But I think for the general public, as long as it functions well, it's going to recede into the background."

If Salvucci is right, then the Big Dig will share the fate of most great achievements in the technological arts. People who use it will take it for granted, and say in passing, "How did we get along without it?" The project's cost is immense. But Americans decided long ago they would pay a lot for good pathways. It's part of the price they pay for their freedom—not only political freedom, but the simple ability to do what they want.

The American Interstate highway system—some forty-two thousand miles in all, the largest road network in the world—likely will rank in history with the roads and viaducts of the Roman Empire. The Big Dig is a fitting final link in that system. For if America is an empire, as its critics say, then at least the Big Dig suggests that an underlying strength of a democratic empire is a willingness to admit and fix its mistakes. For a time, Bostonians got their pathways wrong. So they decided to fix them.

A computer model of the Charles River Bridge. It will be the first "hybrid" cable-stayed bridge in the United States, using both concrete and steel in its frame.

CHAPTER EIGHT

THE INTERNET

Cyberspace. A consensual hallucination experienced daily by billions of legitimate operators, in every nation. . . . A graphic representation of data abstracted from the banks of every computer in the human system. Unthinkable complexity. Lines of light ranged in the nonspace of the mind, clusters and constellations of data. Like city lights, receding . . .

—William Gibson
Neuromancer, 1984

For several weeks in the spring and summer of 1957, an amiable scientist kept a careful record of what he did all day. He was J. C. R. Licklider, a research psychologist at MIT who specialized in psychoacoustics, the study of how the brain perceives sounds and speech. Licklider wanted to know how much time he was devoting to his various tasks. As a scientist, he took it for granted that he spent most of his work time engaged in an activity that he and everyone else would normally call "thinking." But when he kept a record, he found that 85 percent of the time he spent on "thinking" actually went to "getting into a position to think, to make a decision, to learn something I needed to know." He spent a lot of time gathering data, then plotting the data in graphs by hand. One day, for instance, he wanted to compare six different sets of findings on the brain's ability to distinguish speech from mere noise. It took him several hours just to get all the sets of data into a common form that allowed him to make a comparison. Once that was done, his own "thinking"—the comparison itself—took only seconds. It was finding and shuffling the data that took so much time.

Licklider's little experiment confirmed an idea he had harbored for some time. The idea was this: If a machine could find and sort his data, then he and the machine together could do a great deal of thinking.

As early as the 1820s, Englishman Charles Babbage, after spending many tedious hours checking and correcting hand-calculated astronomical tables, had declared, "I wish to God these calculations had been executed by steam!" Babbage had gone on to invent one of the first devices identifiable as a computer. But J. C. R. Licklider lived in a time—the mid–twentieth century—when his vision of people working in harness with computers could lead not just to one new tool but to a great transformation in the way people communicate, think, work, and create.

If the act of invention means to conceive, design, and build a new device, then Licklider did not invent the Internet. He was no Edison, and the Internet is not a monolithic project, not a Pearl Street Station or a magnificent dam or a monumental bridge. It is a vast patchwork quilt of ideas and practices, stitched together and added to through the years, an immense agglomeration of solitary labors and cooperative endeavors. In fact, the Internet never could have emerged in a society where knowledge and power were concentrated. It came into being in a society where knowledge and power resided in thousands of uncelebrated places—laboratories, classrooms, library cubicles, offices, basements, coffee shops, bars—any place where someone who knew about computers and telephone lines could say: "Knowing this, let's try that."

But if Licklider was not the Internet's Edison, he at least played Jack to the Internet beanstalk. Conceiving the seed of an idea in his mind, he flung it outward, hoping it would grow into something resembling his vision. And it did.

J. C. R. Licklider.

"AS NO HUMAN BRAIN HAS EVER THOUGHT . . ."

Born in St. Louis in 1915, Licklider was the only child of a Baptist minister and a true son of the twenties who loved model airplanes and automobiles. As an undergraduate at Washington University in the early thirties he studied fine arts and chemistry before embracing three concurrent majors—physics, math, and psychology. He chose a career in research psychology, but he kept his fingers in harder sciences. At the University of Rochester, he did his doctoral research in neuroscience, the study of the dense web of electronic pathways that make up the brain. He became interested in human perception, specifically in the puzzle of sound localization—how the auditory system determines the location and distance of a given sound. Licklider drew the first maps of the neural wiring in the auditory cortex, the part of the brain that interprets sound. It may be no coincidence that psychoacoustics bred several pioneers in computer networking. The Internet eventually would resemble nothing so much as the intricate, duplicative circuitry of the brain.

During World War II, Licklider worked on military research in

A technician checks for one bad tube, among 19,000 possibilities, in the Electronic Numerical Integrator and Calculator (ENIAC).

the psychoacoustics laboratory at Harvard University. Under an army contract, he was asked to help bomber crews who were straining to discern radio transmissions amid the overpowering noise of wartime cockpits. Licklider devised a way to accentuate consonants and diminish vowel sounds in the transmissions, making speech easier to understand. An observer of Licklider's career points to this as an early instance of his adeptness at shaping technology to fit human needs, rather than demanding that humans bow to technological imperatives.

After the war, Licklider became a professor at Harvard, then at the Lincoln Laboratory at neighboring MIT, which became his intellectual home. His pioneering work in brain science led him to leadership in the emerging field of human factors analysis, the study of human behavior in complex technological systems. He headed human factors work for Project SAGE, the Defense Department's attempt at a computer-based system for defending against Soviet bombers. The SAGE computer was so big the scientists could walk around inside it. But it was one of the first to allow users to interact with it directly via keyboards and get an answer immediately, rather than submitting punchcards that took many hours to process. The SAGE monster caught Licklider's fancy, and he began to ponder the possibilities of human-computer interactions.

Then Licklider met Wesley Clark, a young computer researcher at the Lincoln Lab, who showed him the research capabilities of a computer called TX-2. It was another mammoth, filling two rooms, but it could display graphics on video screens. Licklider was enchanted. The addiction of interactive computing, so rampant in generations to come, took one of its first captives. Licklider soon abandoned psychoacoustics for full-time work in computer science, moving to Bolt Beranek and Newman, a Cambridge consulting firm specializing in acoustical engineering. This was a critical step in the emerging story of computer networking, which is told in depth by Katie Hafner and Matthew Lyon in *Where Wizards Stay Up Late: The Origins of the Internet.*

When Licklider asked Leo Beranek, a former professor of electrical engineering at MIT, to buy a computer for him to use, Beranek asked how much it would cost.

"About twenty-five thousand dollars," Licklider replied.

"That's a lot of money," Beranek said. "What are you going to do with it?"

Lincoln Laboratory, Massachusetts Institute of Technology, 1956.

"I don't know," Licklider said.

Coming from Licklider, that was good enough. BBN was launched in a new direction; the firm would become a powerful force in the development of computer networking, starting with a small working group under Licklider, who soon asked for a still larger investment—a computer built by Digital Equipment Corporation that Licklider bought for nearly one hundred fifty thousand dollars.

SAGE consoles, 1962. The Lincoln Laboratory developed SAGE (Semi-Automatic Ground Environment) out of earlier projects at MIT and IBM.

Licklider was no Einsteinian recluse, thinking deep thoughts in eccentric solitude. Easygoing and generous, he was well-liked by colleagues and students. Indeed, his Middle American amiability led some to underestimate his intellect. "He was the most unlikely 'great man' you could ever encounter," wrote Robert Taylor, a protégé and friend. "His favorite kind of joke was one at his own expense. He was gentle, curious, and outgoing." Like many theorists and scientists, Licklider engaged intellectual problems as a form of sophisticated play, borrowing his game pieces from all the eclectic fields of inquiry

TX-2 at MIT, 1962. The TX-2 took up several rooms and contained 64,000 bytes of memory—about as many as are in a simple handheld calculator in 2001.

he had studied at one time or another. He held ideas up to the light, turning them this way and that, shaking them to guess what was inside, squeezing them into shapes, and putting them to purposes for which they hadn't been intended. "Lick at play with a problem at a briefing or a colloquium, speaking in that soft hillbilly accent, was a *tour de force,*" Bill McGill, a colleague, told Hafner and Lyon. "He'd speak in this Missouri Ozark twang, and if you walked in off the street, you'd wonder, 'Who the hell is this hayseed?' But if you were working on the same problem, and listened to his formulation, listening to him would be like seeing the glow of dawn." Like other great innovators—Edison and the Wright brothers among them—Licklider possessed an extraordinary visual imagination. "He could see the resolution of a technical problem before the rest of us could calculate it," McGill said. "He was like a wide-eyed child going from problem to problem, consumed with curiosity."

That intellectual style allowed Licklider to approach computers with fresh eyes. Others saw only their limitations—the great size and expense, the ponderous process of submitting queries through a screen of technicians via "batches" of keypunch cards. Licklider saw beyond those barriers. The immediate task, he stressed, was to make it possible for individuals to interact with the computer directly and get feedback immediately. The computer must not be a tool only for highly trained technicians. With this in mind, he and his staff attacked the batch computing tradition and developed one of the early

Licklider at MIT's Department of Electrical Engineering in the 1970s.

systems known as "time-sharing" computing, by which several individuals could interact with the mainframe computer directly and at the same time, using separate points of access called terminals. At the same time, he looked much farther ahead. In 1960, he put down his long-range ideas in a fifteen-page article titled "Man-Computer Symbiosis." Published in an obscure technical journal, *IRE Transactions on Human Factors in Electronics,* the article sketched paths of innovation in personal computing and networking that guided researchers for decades to come.

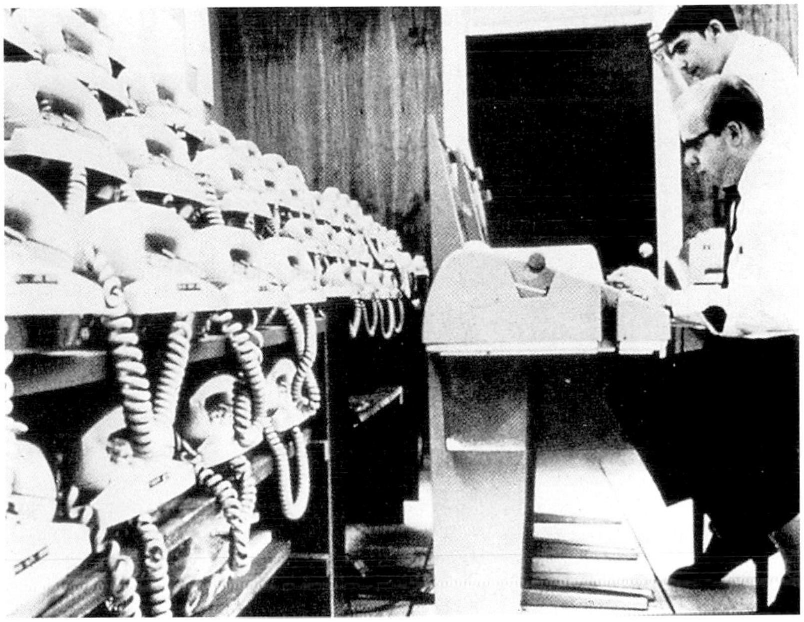

Time-sharing telephones at Bolt Beranek and Newman connected users at individual terminals to a single mainframe host computer.

Licklider began with a consideration of man-machine interactions to date. In some such systems—man and shovel, for example—the tool extends the power of the man. In others—say, a railroad man who shovels coal into a steam locomotive—the man simply fills a role that has not yet been automated. In the case of man and computer, Licklider said, the relationship would become symbiotic: they would thrive upon their mutual dependence. "The hope," he wrote, "is that, in not too many years, human brains and computing machines will be coupled together very tightly, and that the resulting partnership will think as no human brain has ever thought." The years during which this partnership would develop, he said, "should be the most creative and exciting in the history of mankind."

He described the requirements of such a system with uncanny clairvoyance: large viewing screens; vastly enhanced computer memory; programs that would recognize human languages; devices that would allow users to give computers instructions more quickly than by typewriter keystrokes; and fast methods of finding things in a computer's or a network's data banks.

With those tools, he said, a fabulous structure could be built: "It seems reasonable to envision, for a time ten or fifteen years hence, a 'thinking center' that will incorporate the functions of present-day libraries together with anticipated advances in information storage and retrieval and the symbiotic functions suggested earlier. . . . The picture readily enlarges itself into a network of such centers, connected to one another by wide-band communication lines and to individual users by leased-wire services. In such a system, the speed of

"Awake at Last?" by Marcus Edwin, 1957. Uncle Sam, in his bed of "Complacency," is awakened by beeps from the Soviet satellite Sputnik. Computer research and development was a major beneficiary of urgent demands that the U.S. catch up with the Soviets in space.

the computers would be balanced, and the cost of the gigantic memories and the sophisticated programs would be divided by the number of users."

To reassure squeamish colleagues, and perhaps himself, Licklider hastened to say that books would live on in his brave new world: "Books are among the most beautifully engineered . . . components in existence, and they will continue to be functionally important within the context of man-computer symbiosis. (Hopefully, the computer will expedite the finding, delivering and returning of books.)"

If "Man-Computer Symbiosis" and the essays that followed had been Licklider's only contributions to computer networking, he would be remembered merely as a gifted prophet. But in 1962 he received an extraordinary invitation to give concrete form to his ideas. It came from the Advanced Research Projects Agency (ARPA), an office created in the Defense Department to ensure, in the wake of Sputnik, that the United States would regain its technological edge over the Soviet Union. Jack Ruina, ARPA's director, was interested in using computers to augment the military's command-and-control systems. He urged Licklider to head up that effort as well as ARPA's behavioral sciences program. Reluctant, Licklider agreed only when promised a large measure of freedom, a powerful new computer, and a $10 million annual budget.

Licklider served as director of ARPA's Information Processing Techniques Office for only two years, but in that time he planted the seeds that sprouted into the Internet. He went from campus to campus—Stanford, Berkeley, UCLA, Carnegie Tech (soon to be Carnegie-Mellon), MIT—seeking the best minds in computer science and telling them he meant to fund their fondest dreams. He quashed what he called "asinine kinds of things" dreamed up in the Pentagon—for example, a plan to computerize the day-to-day behavior of Kremlin officials in hopes of divining their war plans—and instead seeded nonclassified research on how computers might become easier to use, more powerful, and capable of communicating with each other. Out of the scientists he recruited, an informal network emerged. Ruina called it "Lick's Priesthood." Licklider himself called it, with tongue in cheek, the "Intergalactic Computer Network." In April 1963, he sent a long memo to his acolytes in which he proposed an actual computer network that would pierce the babble of competing programming languages. "Consider the situation in which several different [computing] centers are netted together, each center being highly individualistic and having its own special language and its own special way of doing things," he wrote. "Is it not desirable or even necessary for all centers to agree upon some language or, at least, upon some conventions for asking such questions as 'What language do you speak?'"

Before that question was answered definitively, Licklider left ARPA in 1964. The tasks he laid out fell to those he had inspired and funded.

THE FIRST NETWORK

Bob Taylor was another preacher's kid, another psychoacoustician, and a devotee of Licklider's vision of computers as tools to enhance human power. In 1966, he took over Licklider's old role at ARPA, where he listened to the nation's leading computer scientists plead for money to buy bigger and better computers. In a move more practical than visionary, Taylor went to physicist Charles Herzfeld, then director of ARPA, and proposed that ARPA set up an electronic network among its contractors, most of them at major universities. Over such a network, researchers could share computing resources—for example, an expensive mainframe devoted to graphics research—and exchange their findings and data. At first, as a test, the network could include just a few centers, Taylor said. If it worked, it would deliver more bang for each research buck. Herzfeld asked if such a network would be hard to build. Taylor said no, they already knew how. "Great idea," Herzfeld said. "Get it going. You've got a million dollars more in your budget right now."

The computer network that became known as ARPANET, the direct ancestor of the Internet, was the product of many innovations, large and small, with many architects contributing their designs. At the beginning, several workers in the field moved toward one of the critical early milestones—the conceptual breakthrough that would become known as packet switching.

At RAND, a citadel of Cold War research, Paul Baran worried about the effects of a Soviet atomic attack on the United States military's communications system, often referred to as the command-and-control system. Traditional communications networks were either centralized, with all lines running outward like spokes from a single hub, or decentralized, as in the United States, where connections ran among many such hubs, with spokes branching outward from each. If a Soviet strike destroyed even a handful of hubs, the command-and-control structure could be devastated. The military's options would be reduced to two—no response at all or an all-out counterstrike. Both superpowers were vulnerable in this way, and Baran feared that the situation increased the danger of a preemptive first strike by leaders fearing they would be rendered deaf, dumb, and blind if the other side struck first. Baran called this "the most dangerous situation that ever existed."

Baran proposed to do something about it. He developed the idea of a "distributed network"—a web of connections with no critical centers but only a scattering of crossroads, or "nodes." If an attack destroyed any one node, signals could be routed around the break. Like a spider's web, a distributed network could survive one broken strand, even several. "A redundancy of connectivity," Baran called it.

Paul Baran. When Baran failed to find a parking spot at UCLA one day, he "concluded that it was God's will that I should discontinue school," and he did just that.

Just as Baran was telling fellow researchers about his ideas and urging AT&T engineers to develop them—a plea they rejected—Leonard Kleinrock was thinking along similar lines. As a graduate student at MIT, Kleinrock wrote a dissertation describing a distributed network and an ingenious system for using it. Put very simply, the idea was to divide each electronic message into pieces, then send each piece along the distributed network by itself. Each piece would take a different route, but all would end up at the same destination, where the pieces would be reassembled into an intelligible whole. At many spots along the various paths of the network, computers called routers would act as traffic cops, storing a message for an infinitesimal instant, judging the best route to the destination, then sending the piece along the next leg. An image arose that would later characterize the Internet: myriad fragments of data shooting in every direction at once, chaos in motion, but conjoining at their destinations in orderly ranks.

This idea became a key element in the ARPANET, though the question of credit is tricky. Paul Baran wasn't the only one to propose the same idea, nor was Kleinrock. A British researcher named Donald Davies developed it too. Baran called the idea "message blocks"; Kleinrock called it "time-slicing"; Davies called it "packet-switching," the name that caught on. But ARPANET's designers became familiar with Kleinrock's formulation first, and Kleinrock became one of those designers. So he is often ascribed honors as the pioneer of packet-switching. Historically, the point is not who gets the ribbon, it's that ARPANET now had the means to create a new kind of electronic network, efficient and robust. As Paul Baran reflected later in an interview with the Babbage Institute, a center for the history of computing, "The process of technological developments is like building a cathedral. Over the course of several hundred years, new people come along and each lays down a block on top of the old foundations, each saying, 'I built a cathedral.' Next month another block is placed atop the previous one. Then comes along a historian who asks, 'Well, who built the cathedral?' Peter added some stones here, and Paul added a few more. If you are not careful, you can con yourself into believing that you did the most important part. But the reality is that . . . everything is tied to everything else."

For the job of organizing, building, and administering ARPANET, Bob Taylor tapped an MIT engineer-scientist not yet thirty years old, Larry Roberts. Roberts was reluctant to leave MIT's Lincoln Lab, but after heavy pressure from his superiors and Taylor, Roberts took the job in late 1966. Though deeply reserved, he had a reputation as a superb scientist and administrator with an overpowering drive to acquire information. Roberts studied and practiced speed-reading until he could scan thirty thousand words a minute,

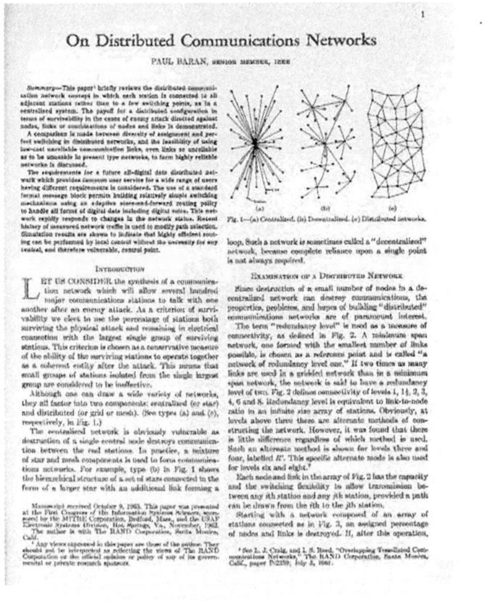
1

On Distributed Communications Networks

PAUL BARAN, SENIOR MEMBER, IEEE

Summary—This paper[1] briefly reviews the distributed communication network concept in which each station is connected to all adjacent stations rather than to a few switching points, as in a centralized system. The payoff for a distributed configuration in terms of survivability in the cases of enemy attack directed against nodes, links or combinations of nodes and links is demonstrated.

A comparison is made between diversity of assignment and perfect switching in distributed networks, and the feasibility of using low-cost unreliable communication links, even links so unreliable as to be unusable in present type networks, to form highly reliable networks is discussed.

The requirements for a future all-digital data distributed network which provides common user service for a wide range of users having different requirements is considered. The use of a standard format message block permits building relatively simple switching mechanisms using an adaptive store-and-forward routing policy to handle all forms of digital data including digital voice. This network rapidly responds to changes in the network status. Recent history of measured network traffic is used to modify path selection. Simulation results are shown to indicate that highly efficient routing can be performed by local control without the necessity for any central, and therefore vulnerable, control point.

Fig. 1—(a) Centralized. (b) Decentralized. (c) Distributed networks.

INTRODUCTION

LET US CONSIDER the synthesis of a communication network which will allow several hundred major communications stations to talk with one another after an enemy attack. As a criterion of survivability we elect to use the percentage of stations both surviving the physical attack and remaining in electrical connection with the largest single group of surviving stations. This criterion is chosen as a conservative measure of the ability of the surviving stations to operate together as a coherent entity after the attack. This means that small groups of stations isolated from the single largest group are considered to be ineffective.

Although one can draw a wide variety of networks, they all factor into two components: centralized (or star) and distributed (or grid or mesh). (See types (a) and (c), respectively, in Fig. 1.)

The centralized network is obviously vulnerable as destruction of a single central node destroys communication between the end stations. In practice, a mixture of star and mesh components is used to form communications networks. For example, type (b) in Fig. 1 shows the hierarchical structure of a set of stars connected in the form of a larger star with an additional link forming a loop. Such a network is sometimes called a "decentralized" network, because complete reliance upon a single point is not always required.

EXAMINATION OF A DISTRIBUTED NETWORK

Since destruction of a small number of nodes in a decentralized network can destroy communications, the properties, problems, and hopes of building "distributed" communications networks are of paramount interest.

The term "redundancy level" is used as a measure of connectivity, as defined in Fig. 2. A minimum span network, one formed with the smallest number of links possible, is chosen as a reference point and is called "a network of redundancy level one." If two times as many links are used in a gridded network than in a minimum span network, the network is said to have a redundancy level of two. Fig. 2 defines connectivity of levels 1, 1½, 2, 3, 4, 6 and 8. Redundancy level is equivalent to link-to-node ratio in an infinite size array of stations. Obviously, at levels above three there are alternate methods of constructing the network. However, it was found that there is little difference regardless of which method is used. Such an alternate method is shown for levels three and four, labelled *R'*. This specific alternate mode is also used for levels six and eight.[2]

Each node and link in the array of Fig. 2 has the capacity and the switching flexibility to allow transmission between any *i*th station and any *j*th station, provided a path can be drawn from the *i*th to the *j*th station.

Starting with a network composed of an array of stations connected as in Fig. 3, an assigned percentage of nodes and links is destroyed. If, after this operation,

Manuscript received October 9, 1963. This paper was presented at the First Congress of the Information Systems Sciences, sponsored by the MITRE Corporation, Bedford, Mass., and the USAF Electronic Systems Division, Hot Springs, Va., November, 1962. The author is with The RAND Corporation, Santa Monica, Calif.

[1] Any views expressed in this paper are those of the author. They should not be interpreted as reflecting the views of The RAND Corporation or the official opinion or policy of any of its governmental or private research sponsors.

[2] See L. J. Craig, and I. S. Reed, "Overlapping Tessellated Communications Networks," The RAND Corporation, Santa Monica, Calif., paper P-2359, July 5, 1961.

"On Data Communications Networks," by Paul Baran, 1964. Baran was among the first to publish research on secure packet-switched networks, the concept that made exchange of information on networked computers possible.

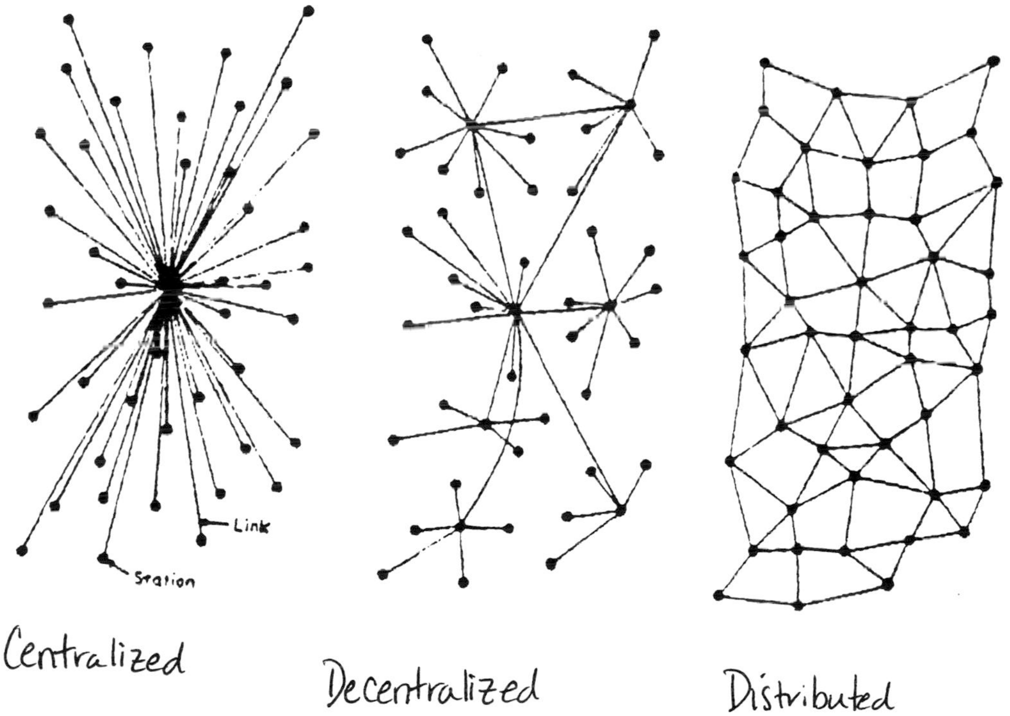

Baran's sketch of (left to right) centralized, decentralized, and distributed networks.

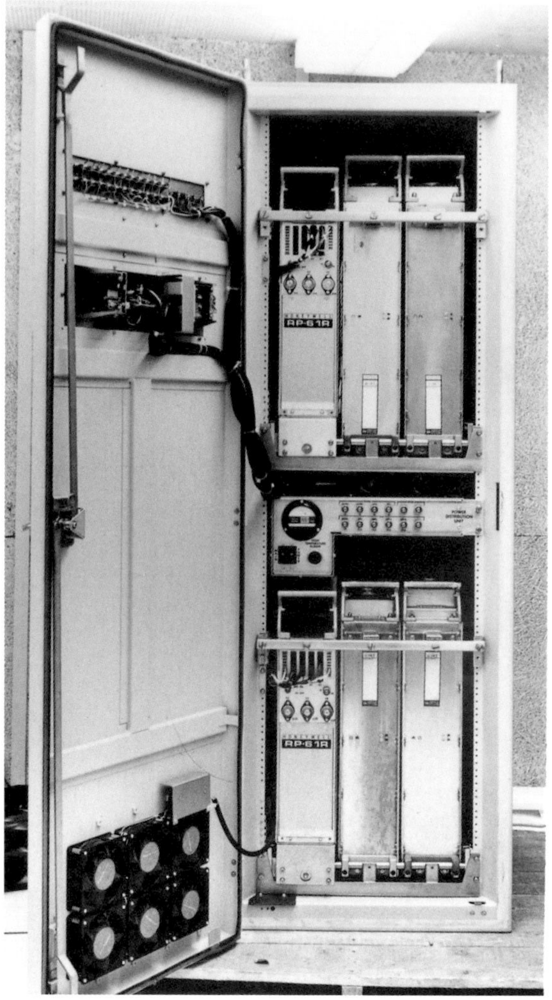

The Interface Message Processor (IMP). It routed information from one computer in a network to another, even if the computers spoke different languages.

with a rate of "selective comprehension" that he estimated at 10 percent. In the vast maze of the Pentagon, Roberts used a stopwatch to determine the quickest routes between offices. According to Hafner and Lyon, "More than a few people had had the experience of explaining to Roberts something they had been working on intensively for years, and finding that within a few minutes he had grasped it, turned it around in his head a couple of times, and offered trenchant comments of his own."

Roberts set about the task of establishing a working network among the university computers sponsored by ARPA. He was soon grappling with the problem that Licklider had anticipated—how to get different computers speaking different languages to communicate with each other. He also ran up against a more human problem—scientists who, though funded by ARPA, were deeply reluctant to get involved in any networking experiment. They didn't care to contribute precious computer time for it, and they doubted the benefits of sharing data with their counterparts on other campuses. Jon Postel, a researcher who would make key contributions to the Internet, told Hafner and Lyon: "People were thinking, 'Why would I need anyone else's computer when I've got everything right here? What would they have that I want, and what would I have that I want anyone else to look at?'"

At the suggestion of Wesley Clark, the MIT researcher who had introduced Licklider to the world of computers, Roberts pursued a scheme that might resolve at least the first of these objections—the problem of communication between host computers that spoke different languages. Instead of creating direct connections via telephone lines among the main computers (called hosts) in the network, ARPA would install a small computer at each host site to handle the administrative work of the network. The small machines would speak the same coding language. Each would act as an intermediary and translator between its host and the wider network.

The job of creating these crucial go-betweens went to Bolt Beranek and Newman, the Cambridge firm that had followed J. C. R. Licklider's lead to become perhaps the leading private center of top-drawer computer talent. At BBN, engineers under Frank Heart, another alumnus of MIT and Lincoln Lab, redesigned Honeywell machines to make the go-between computers, which Larry Roberts had dubbed "interface message processors," or IMPs. (The Internet's addiction to clever coinages predated the Internet itself.) By the fall of 1969, these workhorses, each the size of a refrigerator and weighing nearly half a ton, made it possible for ARPANET to become more than a name. Thomas Hughes has suggested that the achievement of BBN's "IMP Guys" is comparable to that of Edison's electric lighting crew at Menlo Park.

Long strides had been made in the five years since Licklider had left ARPA. Yet in later years Larry Roberts continued to give Licklider much credit for the original conception of computer networking. "The vision was really Lick's originally," Roberts told the makers of the PBS documentary *Nerds.* "None of us can really claim to have seen that before him, nor [can] anybody in the world. . . . He didn't have a clue how to build it. He didn't have any idea how to make this happen. But he knew it was important, so he sat down with me and really convinced me that it was important and convinced me into making it happen."

Leonard Kleinrock and the Interface Message Processor.

Plans were laid in 1968 for the actual, working ARPANET to begin at four academic computing sites—UCLA, the Stanford Research Institute in Menlo Park, the University of California at Santa Barbara, and the University of Utah. But more work was needed before their host computers could shake hands and communicate. A set of coded rules or standards was needed that all hosts would abide by. "It's like picking up the phone and calling France," Frank Heart told a researcher. "If you don't speak French you've got a little problem."

A handful of graduate students at the early ARPANET sites wrote these go-between codes. Two stars of this group were Vinton Cerf and Steve Crocker, old high school buddies from Van Nuys, California. Cerf and Crocker had become fascinated by computers together; once they had sneaked through an upper-story window of the UCLA computing center on a Saturday to gain unauthorized access to its computer. Now they were reunited at UCLA under Leonard Kleinrock, who was overseeing an ARPA-funded program to test and analyze the network. The students were probing an uncharted frontier. "We were just rank amateurs," Cerf later told an interviewer. "We were expecting that some authority would finally come along and say, 'Here's how we are going to do it.' And nobody ever came along."

Cerf, Crocker, and their counterparts at other campuses became known as the Network Working Group. Hafner and Lyon call them "an adhocracy of intensely creative, sleep-deprived, idiosyncratic, well-meaning computer geniuses." Crocker helped to set the group's friendly tone with the minutes he began to keep of their meetings. He titled each of these documents "Request for Comments," a rhetorical device that encouraged participants to feel their ideas were welcome, the process fluid, the decision-making democratic and consensual. In fact, the group's harmonious relations are suggested by the label they chose for the communications codes they were writing—"protocols," a word for the courtly etiquette of diplomacy among sovereign nations.

Vint Cerf, one of the authors of the computer code that allowed computers to talk to each other.

The first IMP was installed at UCLA early in September 1969, the second a month later at the Stanford Research Institute (SRI), several hundred miles to the north. Each IMP was connected to its host computer. But a true network would be established only when the two hosts talked to each other through their respective IMPs. It was agreed that the experimenters at UCLA would have their host computer send a command to the Stanford host. The command would be the term "L-O-G-I-N," meaning that the UCLA people wanted to start a session on the Stanford computer. Hafner and Lyon tell what happened next:

"Fastened to the first IMPs like a barnacle was a small phonelike box, with a cord and headset. It shared the line with the IMPs and used a subchannel intended for voice conversations. The voice line was, like the data line, a dedicated link. A few days after the IMP was in place at SRI, Charley Kline, then an undergraduate at UCLA, picked up the telephone headset in L.A. and pressed a button that rang a bell on the IMP in Menlo Park. A researcher . . . at SRI answered it. It was somehow more thrilling to Kline than dialing a regular telephone.

"The quality of the connection was not very good, and both men were sitting in noisy computer rooms, which didn't help. So Kline fairly yelled into the mouthpiece: 'I'm going to type an *L*!' Kline typed an *L*.

'Did you get the *L*?' he asked. 'I got one-one-four,' the SRI researcher replied; he was reading off the encoded information in octal, a code using numbers expressed in base 8. When Kline did the conversion, he saw it was indeed an *L* that had been transmitted. He typed an *O*.

'Did you get the *O*?' he asked.

'I got one-one-seven,' came the reply. It was an *O*.

Kline typed a *G*.

'The computer just crashed,' said the person at SRI."

The first computer network had been born. And the first network bug had struck. But the bug was quickly exterminated, and by January 1970 ARPANET extended to UC–Santa Barbara and the University of Utah. A long and arduous process of experimentation and testing ensued as researchers filled the network's cables with tiny "packets" of information.

By 1971, Larry Roberts believed the network needed a public unveiling to gain interest and support. A demonstration was planned for October 1972, when the first International Conference on Computer Communication would be held at the Washington, D.C., Hilton. The organizers decided it would be time to put up or shut up; they would allow people to sit at terminals of different makes and actually communicate with host computers throughout the network—twenty-nine of them by that fall. Congressmen, Pentagon officials, and major business leaders would be invited. Computer science teams in academe and business hustled to get a piece of the action, and the conference deadline accelerated the work of finalizing protocols and improving the system. And planners set up things for participants to actually *do* on ARPANET—various games, an interactive quiz about South American geography, and an air-traffic-control simulation.

At last the participants converged on a large meeting hall at the hotel, and a riot of cable-stringing began. Robert Kahn, a key player, recalled: "If somebody had dropped a bomb on the Washington Hilton, it would have destroyed almost all of the networking community in the U.S. at that point." But no bombs fell, the connections worked, and the conference was a remarkable success. Most important, people outside ARPANET saw that it worked. "It was almost like the rail industry disbelieving that airplanes could really fly until they actually saw one in flight," Robert Kahn said.

The Washington conference was a milestone in ARPANET's rise to maturity. But other networks had arisen, too. Network connections were being established by radio, even by satellite, and not only in the United States but in Europe. Yet each network was a world unto itself. To several ARPANET pioneers, the next great challenge was becoming obvious. To realize the technology's potential, there must be a network of networks. That would depend on a new set of internationally accepted protocols—gateway codes that every independent network could use to communicate with all other networks.

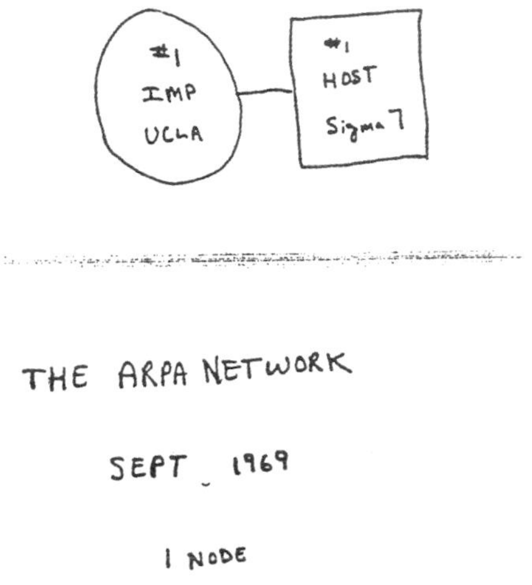

The ARPA network was launched in September 1969. The first "node" was at UCLA (above). History was made by connecting to the site at Stanford Research Institute on October 25. By December, the third and fourth nodes had been added (below).

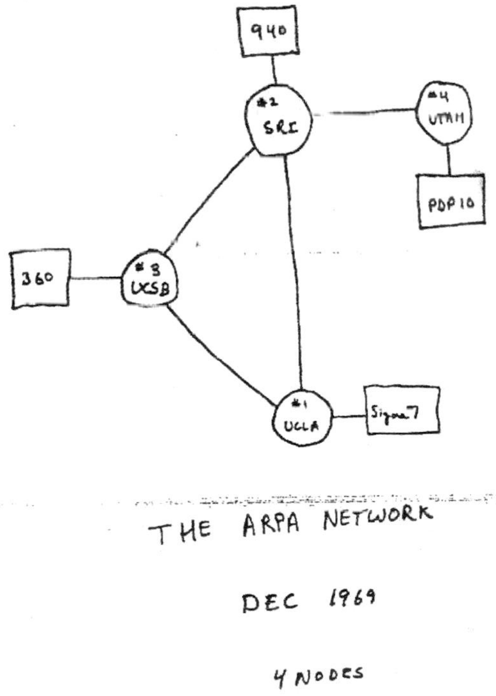

Robert Kahn, then at ARPANET in Washington, and Vinton Cerf, then at the Stanford Research Institute, began to collaborate on the solution in 1973. They called their handiwork the Transmission Control Protocol (TCP). Under TCP, a new set of gateway computers linked to independent networks would insert messages into electronic containers, just as a gift is inserted in a box for shipping. Like shippers, TCP wouldn't care what was in the box; it just needed the boxes to be of uniform size and clearly labeled. So each box would be addressed in a standard code that all the gateway computers could understand. That way, each box could make its way from one independent network to another, and eventually to the intended recipient.

By 1977, Cerf, Kahn, and several colleagues were able to send a message from San Francisco to London via three networks using radio, telephone wires, and satellites. In 1978, a major refinement

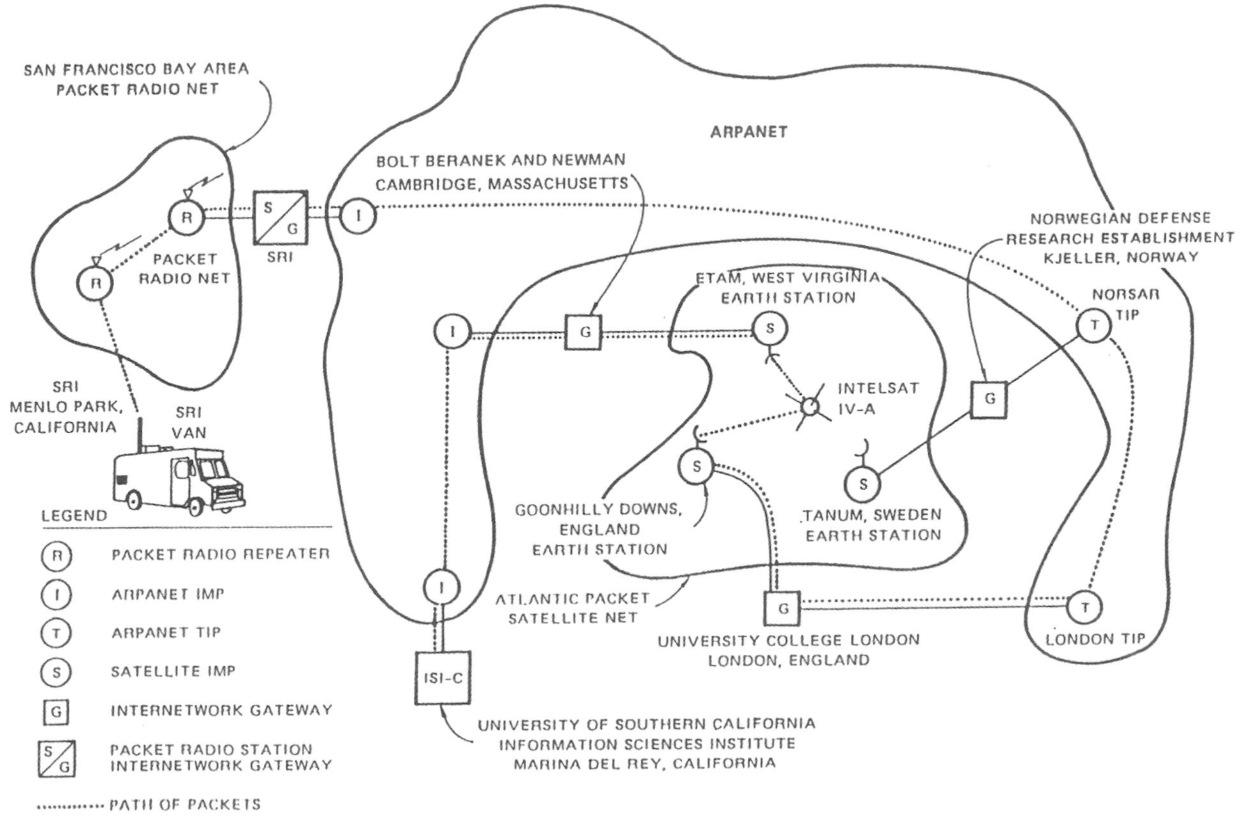

In 1977, Vint Cerf and Robert Kahn mounted a major demonstration, 'internetting' between three networks—the Packet Radio net, SATNET, and the ARPANET. Messages went from a van in the Bay Area across the U.S. on ARPANET, then to University College London and back via satellite to Virginia, and back through the ARPANET to the University of California's Information Services Institute. The demonstration showed how networks could be used internationally.

was added. TCP would be used at each end of an inter-network communication—at the starting point to break the message into packets, and at the destination to put the packets back together in their proper order. In between, a new code called the Internet Protocol (IP) would handle the traffic-cop role of sending the packets along the most efficient routes. Together, TCP/IP became the key that unlocked the potential Licklider had foreseen twenty years earlier. The codes made it possible to connect networks all over the world.

ELECTRONIC COMMUNITIES

The creation of the TCP/IP protocols was perhaps the crowning feat of engineering among all the feats that made the Internet technically possible. But as Hafner and Lyon write, "The romance of the Net came not from how it was built or how it worked but from how it was used." That romance began with simple pieces of electronic mail, the tool that soon convinced dubious academics around the country that a network of computers was a very cool idea.

Though the intended purpose of ARPANET (beyond proving that a computer network was possible) was to share computer resources among the network sites, users had begun sending each other messages almost as soon as the network was in place. But the procedure was laborious. Then, in 1972, Ray Tomlinson, an engineer at Bolt Beranek and Newman, wrote a pair of computer programs that made it not just possible but easy for someone at one computer to send a typed message to someone at another computer far away. One of Tomlinson's innovations, simple but crucial, was a code by which a message could be sent to a particular user at a computer host shared by many people. As he was writing the code, he needed a symbol linking the user's name with the name of the computer host. He looked over his model 33 Teletype keyboard and chose the @ sign.

Tomlinson's little on-line postal system quickly became highly popular on ARPANET. Officials sanctioned it for official purposes only, but before long, faculty members, engineers, programmers, and especially graduate students were using it for nonofficial communication as well. They called it network mail, the early term for what would become known universally as e-mail. They sent messages one to another and devised a means for sharing messages among members of a group. This led to the first computer conferences—long-distance group discussions of topics of interest, held under the banner of an experimental club on ARPANET called

URIs, HTTP, and HTML," Berners-Lee wrote. "There was no central computer 'controlling' the Web, no single network on which these protocols worked, not even an organization anywhere that 'ran' the Web. The Web was not a physical 'thing' that existed in a certain place. It was a 'space' in which information could exist." Nor did the Web become a commodity. The inventor simply sent the programs out on the Internet for free, asking for nothing more than comments about how they might be improved. From there the Web spread like a beneficial virus until it overlay nearly the entire structure of the Internet.

From the beginning, Berners-Lee envisioned the democratic principles that would govern his creation. He incorporated them into his designs and fought to preserve them as the Web spread. One was universality—the idea that the system should reach wherever anyone wanted it, without barriers and without breaking down into separate, competing spheres. Another principle Berners-Lee called equality—the notion that "a person should be able to link with equal ease to any document wherever it happened to be stored." And he wanted the Web to be decentralized. There would be no clearinghouse of documents, no organization in charge.

E-mail and file transfers were person-to-person exchanges, and the data was inaccessible to anyone else who might want it. The Web, by contrast, was a library where documents could remain in semipermanence, with an infinite capacity for users to add new shelves and new volumes. The library, of course, was only as good as its contents.

By 1994, Berners-Lee had become frustrated with his superiors at CERN, who refused to develop the fledgling Web without major funding from outside sources. "This mindset," he writes, "was disappointingly different from the more American entrepreneurial attitude of developing something in the garage for fun and worrying about funding it when it worked!" So when Michael Dertouzos invited Berners-Lee to move to the United States to create and lead a nonprofit World Wide Web Consortium at MIT, he accepted. He arrived in the United States at the beginning of the great Internet gold rush. "As technologists and entrepreneurs were launching or merging companies to exploit the Web," Dertouzos wrote, "they seemed fixated on one question: 'How can I make the Web mine?' Meanwhile, Tim was asking, 'How can I make the Web yours?'"

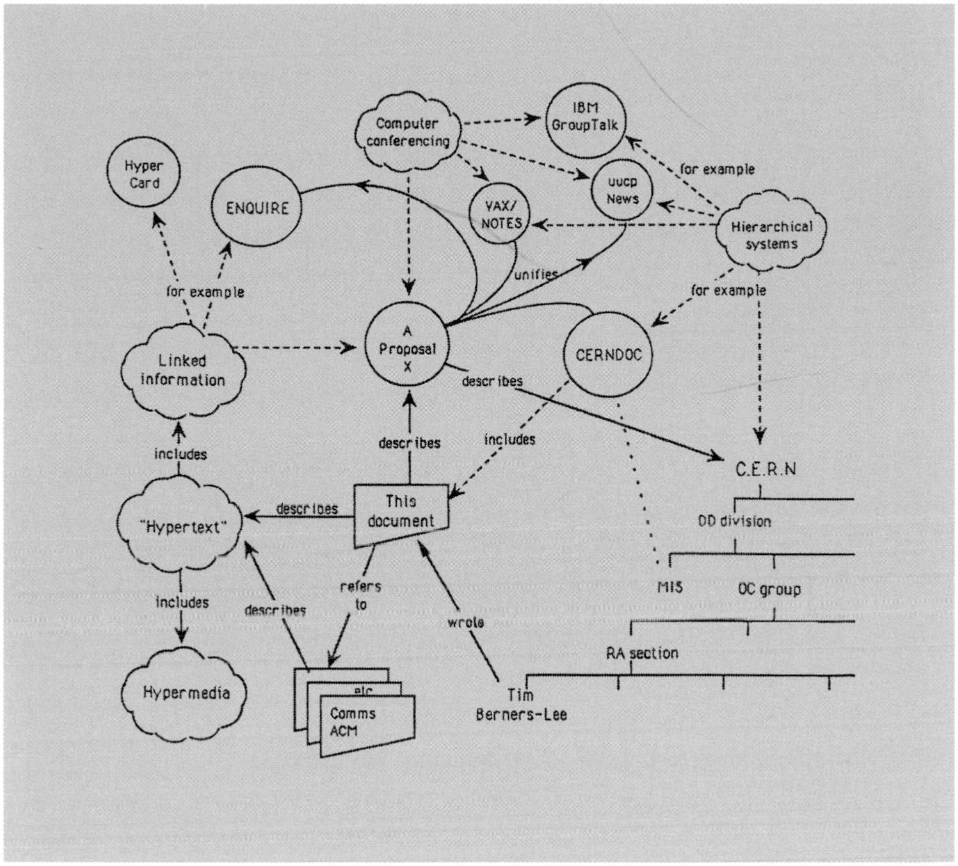

Berners-Lee's proposal for a hypertext information system at CERN. He believed hypertext links would accommodate the laboratory's constant turnover of personnel and information better than a hierarchical or keyword system.

At some point Berners-Lee could have licensed the technology of the Web, formed a company, and capitalized mightily. He had thought about it. But, he wrote, "my motivation was to make sure that the Web became what I'd originally intended it to be—a universal medium for sharing information. Starting a company would not have done much to further this goal, and it would have risked the prompting of competition, which could have turned the Web into a bunch of proprietary products." Though he had designed a program to retrieve and display hypertext documents—a "browser"—he had chosen not to refine it. His decision left an open field for Marc Andreesen, a student at the University of Illinois, who developed an easy-to-use browser called Mosaic, which decoded HTML into instructions for making all the content in a Web document—text, colors, images, sounds—land in the right place and make sense. Andreesen's creation was fundamental to the Web's success. It was also the beginning of Netscape, the company that made the Internet accessible to millions and then went to war with the Microsoft empire, maker of its own competing browser.

In Berners-Lee the world was lucky to get an inventor with an extraordinarily collaborative and generous turn of mind. As he tries to alleviate the confusion that many computer users feel as they adjust to the great innovations of the 1980s and '90s, he expresses his innate sense of the connections between things. "The people of the Internet built the Web, in true grass-roots fashion," he writes in *Weaving the Web.* "On the Net, the connections are cables between computers. On the Web, connections are hypertext links. The Web exists because of programs that communicate between computers on the Net. The Web could not be without the Net. The Web made the Net useful because people are really interested in information (not to mention knowledge and wisdom!) and don't really want to have to know about computers and cables."

For all the millions made off the Internet, most of its pioneers remained working scientists and engineers, prominent in their fields but hardly rich and famous. After his founding stint at ARPA, J. C. R. Licklider worked at Bolt Beranek and Newman, MIT, IBM, and again at ARPA, but his critical contributions were complete. Parkinson's disease and severe asthma sapped his strength. He died in 1990 at seventy-five, much praised by colleagues but mostly forgotten by the wider world.

Robert Taylor, Larry Roberts, Leonard Kleinrock, Paul Baran, Vinton Cerf, Robert Kahn, Doug Englebart all remained in their fields, moving between the private and public sectors, helping the Internet become a phenomenon much closer to Licklider's vision of "man-computer symbiosis" than anyone who broke in on the ponderous, room-sized behemoths of the 1960s had dared to imagine.

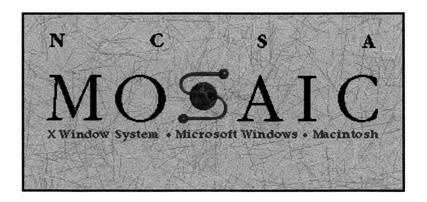

MOSAIC logo. In 1993, students at the University of Illinois created MOSAIC, the world's first Web browser.

Time magazine cover, January 1, 1983. Throughout 1982, _Time_ charted the "computer revolution," devoting three covers to the topic and introducing a regular "Computers" section in April. In this issue, _Time_ wrote, "Since no one person dominated this process, _Time_'s Man of the Year is not a man but the computer itself."

AFTERWORD

Why have Americans built so well?

All civilizations have built great works of one kind or another; indeed, American builders have borrowed from many of them, from the first efforts to control the Mississippi down to the importation of European tunneling techniques in twenty-first-century Boston. Yet Americans have been particularly eager to attempt great works of public engineering, and particularly good at achieving them. And the stories in this book at least suggest tentative ideas about why that is so.

One set of factors lies, of course, in the nation's powerful economy, which has provided the private and public capital essential to major engineering projects, and the incentive for private corporations to take part. Capital accounts for the *existence* of these great projects, but their democratic *character*—the emphasis on spreading technology's benefits to a widening circle of beneficiaries—owes more to the power of the American mass market, a vast middle class of relatively affluent consumers whose dollars could support massive enterprises. The projects have been developed to serve the society as a whole because most members of the society could afford them. The Reclamation Bureau's decision to make electricity with a great dam in the West; New York's decision to import clean water; Thomas Edison's and Samuel Insull's vision of universal electrification; Othmar Ammann's conception of the George Washington Bridge as a bridge for automobiles; J. C. R. Licklider's prophecy of a great network of computers—all these made sense only in a society where most people could afford to pay for water, electricity, cars, and computers.

A certain view of nature arose among a people descended from settlers in a wilderness, and this, too, has fostered large ambitions in engineering. From the first colonists, Americans embraced the task of turning wilderness to their advantage, so they came to regard nature as a thing to be mastered. Certainly this notion was woven through efforts to control the nation's rivers. It was evident, too, in Edison's drive to master electricity, and in the great bridge-builders' approach to rivers.

Tocquevillian individualism has played a role. Another Frenchman schooled in the United States—Robert Calliau, the friend and collaborator of Tim Berners-Lee—has said that all good technology begins with one individual's curiosity. And it's fair to say that individual curiosity, nurtured by a tradition of public education and an admiration for lonely tinkerers, has enjoyed a broad field of play in the United States. Individualism has entailed a defiance of seniority rights, too. Henry Shreve, James Eads, Edison, Insull, John Jervis, Ammann, Fred Salvucci, and most of the Internet pioneers all began to do big things as young men. And that tradition of opportunity for the young acted as a magnet on the Europeans who figure in these tales—Insull, Gustav Lindenthal, Ammann, and Berners-Lee. All these figures prospered. But their deepest drives were not toward wealth or fame. They were possessed by a zeal for creation, by an impulse to make the pictures in their own minds real.

Individualism notwithstanding, these projects were, of course, huge cooperative endeavors, both public and private. Bitter political battles have been waged over which sector should take the lead in building such projects. Yet through the course of this debate, over many decades, a tradition of public/private cooperation has emerged, often fractious and tense but also highly productive and beneficial. The Croton Aqueduct, Hoover Dam, the George Washington Bridge, the Big Dig, the Internet—none are conceivable as projects of private enterprise alone. Nor could any have been built by government alone. Edison and Insull brought electricity to a vast number of households in the pursuit of profit. Yet Insull himself sought public regulation of the utility industry, and without the Tennessee Valley Authority, the Rural Electrification Administration, and other public and quasipublic bodies, the electric lines would have stopped short of supplying the entire country.

The culture of democracy has wielded a subtle but powerful influence in the spread of American technology. It is generally assumed that good things should not be only for elites but for everybody. Reality falls short of that ideal in the lives of many people. But the ideal has power in American society, and that power pushes technological achievements outward from the sources of creativity to broader and broader circles of people, offering them in turn greater measures of opportunity and freedom.

This book has not explored the problems that technology brings in its wake. When we look upon a spectacle such as Boston's Big Dig, in which ever more sophisticated and expensive technology is brought to bear on the problems caused by earlier technology, it all can seem like a futile round-robin, a puppy chasing its tail. Sometimes it's enough to make us long for what is invariably called "a simpler time."

Would we want to go back? We *think* we would. How enchanting it would be to travel back in time to, say, New York City in 1835. There we would discover our ancestors facing the same problem we often face: how to create a new technology (in their case, the Croton Aqueduct) to solve problems caused by older technologies (the canals, railroads, and sail power that had caused the city to prosper and grow). And after a few hours of wonder at a world without clean running water, decent roads, or speedy communications, the enchantment likely would fade. Surely we soon would find ourselves suggesting a better way to do this or that—or, more likely, we would seek out people with that special turn of mind that leads first to rough sketches, then to big plans, and finally to great projects.

AUTHOR'S NOTE ON SOURCES

The narratives herein are based on the research and narratives of Great Projects Film Co. and on the work of scholars, journalists, and researchers who studied the original records and conducted interviews in order to write detailed, authoritative studies. Readers who want more information about these projects are directed to the works listed below for each chapter.

Two books by the eminent historian of technology Thomas P. Hughes have informed the work as a whole. They are *American Genesis: A Century of Invention and Technological Enthusiasm, 1870–1970* (Viking, 1989) and *Rescuing Prometheus* (Pantheon, 1998). Perhaps more than any other recent scholar, Hughes has shown that the tapestry of American history is incomplete without the strands of invention, engineering, and technology. For ideas about the relationship between technology and the culture of democracy in the United States, I am indebted to Daniel Boorstin's *The Americans: The Democratic Experience* (Random House, 1973) and *Democracy and Its Discontents* (Random House, 1974). Ellis L. Armstrong, ed., *History of Public Works in the United States, 1776–1976* (American Public Works Association, 1976), is a solid reference that assisted me at many points.

The passage in the Introduction from the script of *It's a Wonderful Life* is taken from Jeanine Basinger, *The It's a Wonderful Life Book* (Knopf, 1986).

Chapter One: **The Lower Mississippi**

The most important source for the chapter is John M. Barry, *Rising Tide: The Great Mississippi Flood of 1927 and How It Changed America* (Simon & Schuster, 1997). Much more than an account of the flood alone, Barry's book is a narrative of the entire battle over how best to control the Mississippi, a study of Southern culture, and a fine example of the narrative nonfiction genre. Pete Daniel, *Deep'n As It Come: The 1927 Mississippi River Flood* (Oxford University Press, 1977), provided further details on the flood, as did W. A. Percy's memoir, *Lanterns on the Levee: Recollections of a Planter's Son* (Knopf, 1941).

Francis Dorsey's lively biographies, *Master of the Mississippi: Henry Shreve and the Conquest of the Mississippi* (Houghton Mifflin Co., 1941) and *Road to the Sea: The Story of James B. Eads and the Mississippi River* (Rinehart & Co., 1947), provided a number of vivid details. Edith McCall, *Conquering the Rivers: Henry Miller Shreve and the Navigation of America's Inland Waterways* (Louisiana State University Press, 1984), is the fullest account of Shreve's life. A contemporary work by an Eads assistant, E. L. Corthell, *A History of the Jetties at the Mouth of the Mississippi River* (John Wiley & Sons, 1880), is the fullest account of that great project. Several surveys written in a fading tradition of regional history remain interesting: Hodding Carter, *The Lower Mississippi* (Farrar & Rinehart, 1942); Marquis Childs, *Mighty Mississippi: Biography of a River* (Ticknor & Fields, 1982); and Ray Samuel, *Tales of the Mississippi* (Hastings House, 1955).

The U.S. Army Corps of Engineers' role in the control of the river is covered in Floyd M. Clay, *History of Navigation on the Lower Mississippi* (U.S. Army Engineer Water Resources Support Center, 1983); Albert E. Cowdrey, *Land's End: A History of the New Orleans District, U.S. Army Corps of Engineers* (U.S. Army Corps of Engineers, 1977); Gary B. Mills, *Of Men and Rivers: The Story of the Vicksburg District* (U.S. Army Corps of Engineers, 1978); and Michael C. Robinson, "A Tale of Two Rivers: Harnessing the Colorado and Mississippi, 1928–40" (address to the Symposium on the 50th Anniversary of Hoover Dam, 1985). The account of the Corps' effort to keep the Mississippi in its current bed is derived chiefly from John McPhee's "Atchafalaya" in *The Control of Nature* (Farrar, Straus & Giroux, 1989). Also helpful on the post-1927 period were Edwin Jadwin, "The Plan for Flood Control of the Mississippi River in Its Alluvial Valley," *Annals of the American Academy of Political and Social Science* (January 1928); and Cory Dean, "Corps of Engineers Struggles to Alter Mississippi's Fate," *New York Times*, January 15, 1991.

The descriptions of levee construction and

levee-camp culture are based on "Levees of the Mississippi," *Scribner's,* 1881; "The Mississippi River Problem," *Harper's Monthly,* September 1882; Harry Dickson, "The Fight for the Levees," *Harper's Weekly,* April 11, 1903; and Alan Lomax, *The Land Where the Blues Began* (Pantheon, 1993).

Chapter Two: **The Colorado**

The crucial source here is Joseph E. Stevens's *Hoover Dam: An American Adventure* (University of Oklahoma Press, 1988), a first-class history of the control of the Colorado with detailed accounts of the political, business, and technical developments that led to the dam's creation. Other useful works on the dam and its history are Paul L. Kleinsorge, *The Boulder Canyon Project: Historical and Economic Aspects* (Stanford University Press, 1941), and the Bureau of Reclamation's undated pamphlet *Hoover Dam.* Marion V. Allen, *Hoover Dam & Boulder City* (CP Printing & Publishing, 1983), is an informal but interesting book by a veteran of the dam's construction, part memoir and part oral history. Donald Worster's *Rivers of Empire: Water, Aridity and the Growth of the American West* (Pantheon, 1985) is an authoritative and readable history of the nation's painful efforts to make its desert bloom.

The account of John Wesley Powell and the climate of the Southwest relies on Wallace Stegner's classic biography *Beyond the Hundredth Meridian: John Wesley Powell and the Second Opening of the American West* (Houghton Mifflin Co., 1953); Ann Gaines, *John Wesley Powell and the Great Surveys of the American West* (Chelsea House, 1991); Eugene W. Hollon, *The Great American Desert* (Oxford University Press, 1966); Frank Waters, *The Colorado* (Rinehart and Co., 1946); Walter Prescott Webb, "The American West, Perpetual Mirage," *Harper's,* May 1957; Michael C. Robinson, *Water for the West: The Bureau of Reclamation, 1902–1977* (Public Works Historical Society, 1979); and Henry Nash Smith, *Virgin Land: The American West as Symbol and Myth* (Harvard University Press, 1950). Readers interested in Powell should consult Donald Worster's major biography, *A River Running West: The Life of John Wesley Powell* (Oxford University Press, 2001), which was published just as this book was about to go to press.

For the story of the California Development Company's spectacular failure in the Imperial Valley and of the engineers who saved the valley, the key source is a collection of essays written or compiled by the key engineer, H. T. Cory—*The Imperial Valley and the Salton Sink* (John J. Newbegin, 1915). A good survey is Helen Hosmer, "Triumph and Failure in the Imperial Valley," in T. H. Watkins, ed., *The Grand Colorado* (American West Publishing Co., 1969). Also helpful are Maury Klein, *The Life and Legend of E. H. Harriman* (University of North Carolina Press, 2000), and George Kennan, *E. H. Harriman* (Houghton Mifflin Co., 1922).

Details on Six Companies, Henry Kaiser, and Frank Crowe come from Mark S. Foster's biography *Henry J. Kaiser: Builder in the Modern American West* (University of Texas Press, 1989); Donald Worster's *Rivers of Empire;* "Francis Trenholm Crowe," *Contractors and Engineers Monthly,* 1946; S. O. Harper et al., "Francis Trenholm Crowe," *Transactions,* American Society of Civil Engineers, 1946; and Leonard Lyons, "Unforgettable Henry J. Kaiser," *Reader's Digest,* April 1968. Norris Hundley, Jr., "The Politics of Reclamation: California, the Federal Government, and the Origins of the Boulder Canyon Act—A Second Look," *California Historical Quarterly* 52, no. 4 (1974), treats the dam's political history. See also Hundley's major works: *Water and the West: The Colorado River Compact and the Politics of Water in the American West* (University of California Press, 1975) and *The Great Thirst: Californians and Water, 1770s–1990s* (University of California Press, 1992).

Material on Herbert Hoover's role in the control of the Mississippi and Colorado Rivers is drawn from John M. Barry's *Rising Tide,* cited earlier; David Burner, *Herbert Hoover: A Public Life* (Knopf, 1979); Richard Norton Smith, *An Uncommon Man: The Triumph of Herbert Hoover* (Simon & Schuster, 1984); and *The Memoirs of Herbert Hoover: The Cabinet and the Presidency, 1920–1933* (Macmillan, 1952).

Chapter Three: **Edison's Light**

Thomas Edison has been the subject of many full-length biographies. This chapter relies on four of them: Paul Israel, *Edison: A Life of Invention* (Wiley, 1998); Neil Baldwin, *Edison: Inventing the Century* (Hyperion, 1995); Robert Conot, *A Streak of Luck* (Seaview Books, 1979); and Matthew Josephson, *Edison* (McGraw-Hill, 1959). Israel, managing editor of the multivolume Thomas Edison Papers at Rutgers University, is one of the world's leading Edison experts; his book is the authoritative scholarly biography. I found Baldwin's book highly readable and the most revealing of Edison the man. Conot is especially good at the hard work of making electricity comprehensible to the lay reader. Thomas P. Hughes's essay, "The Electrification of America: The System Builders," in Terry S. Reynolds, ed., *The Engineer in America: A Historical Anthology from Technology and Culture* (University of Chicago Press, 1991), places Edison in a broad historical context.

For the process of inventing what we now call the lightbulb, the indispensable source is Robert Friedel and Paul Israel, *Edison's Electric Light: Biography of an Invention* (Rutgers University Press, 1986), a detailed and well-illustrated work of technological history. Andre Millard, "Machine Shop Culture and Menlo Park," and David Hounshell, "The Modernity of Menlo Park," in William S. Pretzer, *Working at Inventing: Thomas A. Edison and the Menlo Park Experience* (Henry Ford Museum and Greenfield Village, 1989), cover the special environment in which the electric light was developed. Bern Dibner, "The Beginning of Electricity," in Melvin Kranzberg and Carroll W. Pursell, Jr., eds., *Technology in Western Civilization,* vol. I (Oxford University Press, 1967), is an effective overview. Thomas J. Schlereth, *Victorian America: Transformations in Everyday Life, 1876–1915* (HarperCollins, 1991), is informative on the practicalities of gas and electric lighting.

Chapter Four: **Electric Nation**

The chapter is drawn chiefly from two outstanding works of history: Forrest McDonald, *Insull* (University of Chicago Press, 1962); and Harold L. Platt, *The Electric City: Energy and the Growth of the Chicago Area, 1880–1930* (University of Chicago Press, 1991). Platt's book, though a work of local history, illuminates a wide national canvas. The Great Projects Film Company's interviews with McDonald and Platt were helpful as well. The story of Insull's trial is well told in Francis X. Busch, *Guilty or Not Guilty? An Account of the Trials of the Leo Frank Case, the D.C. Stephenson Case, the Samuel Insull Case, the Alger Hiss Case* (Bobbs-Merrill, 1952). Larry Plachno, ed., *The Memoirs of Samuel Insull* (Transportation Trails, 1992), gives Insull's view of events on the eve of his trial.

The advance of electricity as a social and cultural phenomenon is explored in David E. Nye, *Electrifying America: Social Meanings of a New Technology, 1880–1940* (MIT Press, 1991), and Ronald C. Tobey, *Technology as Freedom: The New Deal and the Electrical Modernization of the American Home* (University of California Press, 1996).

The section on the Tennessee Valley Authority relies on Thomas K. McCraw, *TVA and the Power Fight, 1933–1939* (Lippincott, 1971), a solid and readable survey of the controversies inside and outside the agency. Other sources are Martha Munzer, *Valley of Vision: The TVA Years* (Knopf, 1969); George Norris's memoir, *Fighting Liberal* (Macmillan, 1945); Arthur E. Morgan, *The Making of the TVA* (Prometheus Books, 1974) and *The First Fifty Years: Changed Land, Changed Lives* (Tennessee Valley Authority, 1983). Roy Talbert, Jr., *FDR's Utopian: Arthur Morgan of the TVA* (University of Mississippi Press, 1987), covers the life of that strange and interesting public servant.

Chapter Five: **Water for the City**

An account of the building of the Croton Aqueduct scarcely would have been possible without Gerard T. Koeppel's *Water for Gotham: A History* (Princeton University Press, 2000). Koeppel's work, a fascinating story and a thoroughly researched piece of scholarship, illuminates the aqueduct as a social, political, and technological milestone. A useful supplement, with information about many other U.S. water systems, is Nelson Manfred Blake, *Water for the Cities: A History of the Urban Water Supply Problem in the United States* (Syracuse University Press, 1956).

Information on the life of the aqueduct's chief engineer comes from F. Daniel Larkin, *John B. Jervis: An American Engineering Pioneer* (Iowa State University Press, 1990), and Neal FitzSimons, ed., *The Reminiscences of John B. Jervis, Engineer of the Old Croton* (Syracuse University Press, 1971).

Those interested in the history of New York City are blessed with two fine new sources: Edwin G. Burrows and Mike Wallace, *Gotham: A History of New York City to 1898* (Oxford University Press, 1999), and Ric Burns and James Sanders, *New York: An Illustrated History* (Knopf, 1999). These books were helpful on numerous points, as was Oliver E. Allen, *New York, New York: A History of the World's Most Exhilarating and Challenging City* (Atheneum, 1990). Several points come from an article authored by the city's Department of Environmental Protection, "New York City's Water Supply System," and "The History of Westchester: The First 300 Years," a survey article at the county's official Web site, *www.co.westchester.ny.us/history/.*

The account of the Erie Canal is drawn from Ronald E. Shaw, *Erie Water West: A History of the Erie Canal, 1792–1854* (University of Kentucky Press, 1966), and Harvey Chalmers III, *The Birth of the Erie Canal* (Bookman Associates, 1960).

Chapter Six: **Ammann's Bridge**

This story of the linked careers of Othmar Ammann and Gustav Lindenthal comes mainly from Jameson W. Doig, "Politics and the Engineering Mind: O. H. Ammann and the Hidden Story of the George Washington Bridge," *Yearbook of German-American Studies* 25: 151–99; and from Henry Petroski, *Engineers of Dreams: Great Bridge Builders and the Spanning of America* (Knopf, 1995). Doig is to be thanked especially for mining the archival sources that reveal Ammann's remarkable work as a "political entrepreneur." Doig and David P. Billington join forces in "Ammann's First Bridge: A Study in Engineering, Politics, and Entrepreneurial Behavior," *Technology and Culture* 35: 537–70). Doig places Ammann in the context of the broader history of the Port of New York Authority in *Empire on the Hudson: Entrepreneurial Vision and Political Power at the Port of New York Authority* (Columbia University Press, 2000). Billington's superb *The Tower and the Bridge: The New Art of Structural Engineering* (Basic Books, 1983) explains Ammann's pioneering role in engineering design.

Darl Rastorfer, *Six Bridges: The Legacy of Othmar H. Ammann* (Yale University Press, 2000), is a beautifully illustrated survey of Ammann's career and work; it furnished several key details for this chapter. The journalist Tom Buckley's article, "The Eighth Bridge," *The New Yorker,* January 14, 1991, is a good short history of the Hell Gate Bridge. A much earlier *New Yorker* article, Milton MacKaye, "Poet in Steel" (June 2, 1934), offers a contemporary snapshot of Ammann at work. The account of Gay Talese's meeting with the elderly Ammann comes from Talese's article in the *New York Times,* "City Bridge Creator, 85, Keeps Watchful Eye on His Landmarks" (March 26, 1964).

The Brooklyn Bridge is described from very different angles of vision—the first historical, the second cultural—in David McCullough, *The Great Bridge: The Epic Story of the Building of the Brooklyn Bridge* (Simon & Schuster, 1972), and Alan Trachtenberg, *Brooklyn Bridge: Fact and Symbol* (Oxford University Press, 1965).

On technical matters of bridge-building, these works were useful: Donald C. Jackson, *Great American Bridges and Dams* (The Preservation Press, 1988); David Macaulay, *Building Big* (Houghton Mifflin, 2000); and Michael Morrissey, "How Bridges Work," on Marshall Brain's rich "How Stuff Works" Web site (*www.howstuffworks.com*).

The emergence of the automobile as a major force in American society is documented in two books by the historian James J. Flink: *The Car Culture* (MIT Press, 1975) and *The Automobile Age* (MIT Press, 1988).

Chapter Seven: **The Big Dig**

The chapter owes a large debt to those who graciously consented to be interviewed for the *Great Projects* film on the subject, especially Frederick Salvucci, Louis Silano, Michael Dukakis, Alan Altshuler, David Luberoff, Michael Lewis, and Robert Albee.

An essential source was the monograph *Mega-*

Project: A Political History of Boston's Multibillion Dollar Artery/Tunnel Project (revised edition, Harvard University, April 1996) by David Luberoff and Alan Altshuler, of the A. Alfred Taubman Center for State and Local Government, a part of the John F. Kennedy School of Government at Harvard University. Luberoff, a researcher, and Altshuler—director of the center, former Massachusetts secretary of transportation, and a key player in the events leading up to the Big Dig—provide a comprehensive account of the project's long and tangled political history. It is rare to find such a careful work of scholarship on so contemporary a topic.

Thomas P. Hughes's chapter on the Big Dig in *Rescuing Prometheus* is especially helpful on the problems engineers have faced in managing the complexity of the project in all its facets—technical, political, and social. Alan Lupo, et al., *Rites of Way: The Politics of Transportation in Boston and the U.S. City* (Little Brown, 1971), offers a vivid account of Boston's antihighway movement in the 1960s, including Fred Salvucci's role. Charles Kenney and Robert Turner's *Dukakis: An American Odyssey* (Houghton Mifflin, 1988) is a solid political biography with helpful background on the development of Dukakis's views on transportation.

A good source on the technology of the Big Dig is the project's official Web site, *www.bigdig.com,* a rich collection of narratives, illustrations, and other material. The site's photographs are extraordinary.

Information on the history of the Interstate highway system came chiefly from Phil Patton, *Open Road* (Simon & Schuster, 1986); Richard F. Weingroff, "Creating the Interstate System," *Public Roads* (Summer 1996); and Ellis L. Armstrong, ed., *History of Public Works in the United States, 1776–1976* (American Public Works Association, 1976).

On the Scheme Z controversy, articles by Peter J. Howe of the *Boston Globe* were helpful, as were several *USA Today* articles.

The article on land-making in Boston is derived from Alex Krieger et al., *Mapping Boston* (Muriel G. and Norman B. Leventhal Family Foundation, 1999), especially the chapter by Nancy S. Seasholes, "Gaining Ground: Boston's Topographical Development in Maps." That book is a breathtakingly beautiful collection of maps, illustrations, photographs, and expert commentary. Every city ought to have such a treatment.

Chapter Eight: **The Internet**

The chapter relies heavily on Katie Hafner and Matthew Lyon's richly detailed *Where Wizards Stay Up Late: The Origins of the Internet* (Simon & Schuster, 1996). This is a fine narrative account of ARPANET and how it led to the creation of the Internet, technically detailed enough for cyber-sophisticates yet accessible to the curious general reader. Thomas P. Hughes's overview of ARPANET's history in *Rescuing Prometheus* also was a great help. Hafner's article "The Epic Saga of The Well," *Wired,* May 1997, is the source of the chapter's account of that early online community. Hafner's account is expanded in *The Well: A Story of Love, Death & Real Life in the Seminal Online Community* (Carroll & Graf, 2001), which was published as this book went to press. Additional information about J. C. R. Licklider comes from two of Licklider's own articles, including one written with Robert W. Taylor. These articles—"Man-Computer Symbiosis" and "The Computer as a Communication Device"—are reprinted in *In Memoriam: J. C. R. Licklider, 1915–1990* (Systems Research Center, 1990). Also useful were M. Mitchell Waldrop, "Computing's Johnny Appleseed" [on Licklider], *Technology Review,* January/February 2000; and George Gilder, "Inventing the Internet Again" [on Paul Baran], *Forbes,* June 2, 1997.

The section on Tim Berners-Lee is based on the inventor's own book, *Weaving the Web: The Original Design and Ultimate Destiny of the World Wide Web* (HarperBusiness, 1999, 2000), and on *How the Web Was Born* (Oxford University Press, 2000), by Robert Calliau (Berners-Lee's colleague at CERN) and James Gillies (a science writer). The latter book is a comprehensive history. Berners-Lee's own account is especially valuable for the description of his early ideas about information management and for his vision of the medium's future. Quotations from Michael Dertouzos are drawn from his foreword to Berners-Lee's book. Several Berners-Lee quotations are drawn from James Luh, "Tim Berners-Lee," *Internet World,* January 2000.

Information about the origins and development of Usenet comes from Michael Hauben and Ronda Hauben, *Netizens: On the History and Impact of Usenet and the Internet* (IEEE Computer Society, 1997). This detailed account, drawing on the Haubens's communications with many early Usenet users, also appears at the Web site *www.columbia.edu/~hauben/netbook/*

Vannevar Bush's seminal July 1945 article in *The Atlantic,* "As We May Think," can be found on the World Wide Web at *www.theatlantic.com*

ACKNOWLEDGMENTS

I was an engineer once, a long time ago. Sometimes, I wish I still were. Now I am a documentary filmmaker. When I tell people that, most of them are impressed. If I told them I was an engineer, I doubt they would feel the same way. Yet the citizens of the United States owe far more to engineers than to filmmakers. Each day, we walk through a landscape of engineering solutions without which modern life would be impossible. Certainly filmmaking would be impossible. Yet, sadly, we take most of it for granted.

This effort to tell the story of the engineers who built our nation began more than a decade ago. Our goal was to chronicle, for public television viewers, the impact—most of it beneficial—of great engineering and public works projects throughout the nation's history. We hoped our work would encourage people to value these achievements and to support future investments in public works. We also hoped greater knowledge might reduce the widespread fear of technology. Like all human intervention in the natural world, engineering must be accompanied by foresight and wisdom. Usually, it has been so. And the engineers' story is as heroic as any in American history.

Kenneth Mandel

To produce the PBS series *Great Projects: The Building of America,* we looked for collaborators who shared our sense of the heroism of the nation's engineers and who rejected the current academic fashion of devaluing the role of the individual in history. Our search led us to Dan Miller and Seth Kramer, intelligent young documentary filmmakers who, with vast skill, took complicated stories and transformed them into works of cinematic art.

Engineers are problem-solvers. In editing the four programs in the television series, senior editor Andrew Morreale was our engineer, solving each storytelling problem large and small.

When we met award-winning author James Tobin, we knew we had found the right person to write the companion book. He was able to distill years of research into exciting stories, and when we pointed out oversights or suggested a turn of phrase, he always welcomed the constructive criticism.

Bruce Nichols and his colleagues at The Free Press have handled the project with enthusiasm and intelligence from the first day we met, and have provided all the resources and support necessary to complete a complicated undertaking.

Without our agents, Michael Carlisle and Christy Fletcher, this book never would have happened. Special thanks go to our guiding light, Marly Rusoff, who was enthusiastic when we failed to be and who from the start was our strongest advocate.

Elton Robinson and Chris Robinson used their creative talents to make the book a true pleasure to behold. The whole staff at Great Projects Film Company pitched in, especially Libby Kreutz, Sean Loughlin, Kate Delimitros, and, in the early days, Amy Faust and Judy Toong.

We owe thanks to many people who helped us move this story into print and film. At South Carolina Educational Television, the list includes Polly Kosko, Charles White, Elaine Freeman, Sally Foster, Coby Henneey, and the late Charlie Morris; at PBS, Glenn Marcus, Sandy Heberer, Alyce Myatt, and Pat Mitchell.

Our underwriters helped at every stage and remained patient throughout. They are the U.S. Department of Transportation, the Federal Highway Administration, the Alfred P. Sloan Foundation, the National Science Foundation, AT&T, DPIC Insurance, the Harriman Foundation, the Jacobs Family Fund, the Engineering Foundation, Jacobs Engineering, Deloitte Touche, the Construction Industry Manufacturers Association, the Association of Equipment Distributors, the American Society of Civil Engineers, the U.S. Bureau of Reclamation, Morrison Knudsen, Caterpillar, the CIT Foundation, The Road Information Project, and the Canadian Association of Equipment Distributors.

Among the project's underwriters, special mention is reserved for the most patient and supportive man we know, Arthur Singer of the Sloan Foundation, and for his successor, Doron Weber. At the Engineering Foundation, Dr. Charles Freiman was always encouraging, as were Bill Salmon, Robin Gibbin, and Dr. Robert White at the National Academy of Engineering. The academy provided technical expertise on a regular basis whenever it was requested.

We drew on the expertise of many generous engineers, historians, journalists, and friends. Among them are Joel Moskowitz, Bill Abrams, Bob Ubell, Peter Hall, Nancy Connery, George Tamaro, Gene Fasullo, Frank Lombardi, Tom Kuesel, Wilson Binger, Sam Schwartz, Art Fox, Sam Florman, Henry Petroski, Frank Davidson, Alan Gruber, Richard Striner, Howard Stussman, Thomas Hughes, and Margo Ammann. The series and book were shaped by discussions with David Billington, whose insights inform every part of the project, and Lord Asa Briggs, the renowned British historian.

A handful of individuals were important in ways that would take pages to describe. Sante Esposito, Eddie Mahe, Ann Eppard, and Julie Chlopecki will always have our gratitude for all they did for us.

We had bipartisan support from former Representative Bud Shuster, chairman of the House Transportation Committee, and his Democratic counterpart, James Oberstar, who deserve the nation's thanks for their support of public works.

During the making of *Great Projects* we lost two of our most ardent supporters, Isaac Auerbach and Dr. Mike Robinson. We miss them and wish they could have been around to see the fruits of our labor.

Finally, we thank Janet, Arthur, and Tyler; and Eileen, Kate, Ben, and Jake, who have lived with *Great Projects* for so many years of late nights and long trips.

Kenneth Mandel
Daniel B. Polin

I am grateful to many who helped this book to publication.

At Great Projects Film Co., I found fine collaborators and friends in Daniel Polin and Kenneth Mandel and received much help from their colleagues, especially Seth Kramer, Daniel A. Miller, Sean Loughlin, Libby Kreutz, Sonia Slutsky, and Kate Delimitros. They discovered the stories, unearthed the facts, and laid down the narrative lines. Polin's and Mandel's vision is the book's raison d'être. Their wise suggestions and factual corrections improved the manuscript in countless ways. I hope I justified their trust.

Deborah DeGeorge provided skilled assistance with research. Karl Leif Bates, Dave Farrell, and Randy Milgrom offered helpful comments on various drafts. I'm very grateful to Carol Mann, Christy Fletcher, Marly Rusoff, and Michael Carlisle for their hard work in bringing a complicated deal to birth. We were lucky to have The Free Press take on the book; to have Bruce Nichols's intelligent and good-spirited guidance; and to have the highly competent support of Daniel Freedberg, Edith Lewis, and a terrific design and production team.

I thank these experts, who reviewed chapter drafts: Jameson W. Doig, Katie Hafner, Paul Israel, David Luberoff, Gerard T. Koeppel, Harold L. Platt, Fred Salvucci, and Joseph E. Stevens. Their corrections and suggestions saved me from many errors, though they bear no responsibility for the book's interpretations or for any errors that might remain. Their research, and that of other scholars and journalists, is the book's foundation.

Most of all I am grateful to Leesa Erickson Tobin, who helped with research and met difficulties, small and large, with her usual strength and grace.

James Tobin

ILLUSTRATION CREDITS

The sources for the illustrations in this book are listed by page and position on the page (T top, M middle, B bottom, L left, R right). Original maps by Maryland Mapping & Graphics pages 8, 27, 35, 144, 257, 265.

Chapter One

2–3 Elton Robinson • 4–5 Bureau of Reclamation • 7T R. W. Norton Art Gallery, Shreveport, La. • 7B, 16 Historic New Orleans Collection • 11 U.S. Patent Office • 12 Mississippi Department of Archives and History • 14 ("Filling a Low Spot Near the Rock Levee, 1913" by F. A. Rosselle), 15 Delta State University Archives, Cleveland, Miss. • 19, 23 U.S. Army Corps of Engineers • 25 Old Court House Museum Collection, Vicksburg, Miss. • 26 Memphis/Shelby County Public Library & Information Center, Memphis, Tenn. • 27, 33 National Archives • 29, 30, 32 U.S. Army Corps of Engineers • 36 William Alexander Percy Memorial Library

Chapter Two

38 Smithsonian Institution • 39, 40 National Archives • 42 N & N Heil Collection, Imperial Valley Historical Society • 43 Huntington Library, Pasadena, Calif. • 44T Sherman Library and Gardens • 44B, 46T National Archives • 45 Chris Robinson • 46B, 48, 49 Imperial Irrigation District • 47 California State Library • 50 California State Railroad Museum • 52 Library of Congress • 54, 55 Bureau of Reclamation • 57, 58, 61, 63B Nevada State Museum and Historical Society • 59, 60, 62, 63T, 64–69, 71 Bureau of Reclamation • 70 courtesy of Joseph E. Stevens/Bureau of Reclamation

Chapter Three

72–73 C. Mayhew and R. Simmon, NASA and Goddard Space Flight Center Digital Archive • 75–77, 78T, 78BL, 82–87, 89, 91T, 92, 94B, 95–97, 99L, 100T, 101B, 106, 107L, 108 Edison National Historic Site (page 75, charcoal sketch by James Edward Kelly; page 106, "Leak in the Edison System" by Thomas Worth) • 78BR, 79, 88, 100M, 107R Smithsonian Institution • 80–81, 91B Henry Ford Museum and Greenfield Village • 90B, 94T, 100B, 101T, 103, 105 Consolidated Edison Company of New York • 90T Museum of the City of New York • 98 New York Public Library, Map Division • 99R Thomas A. Edison Papers, Rutgers University • 102 Brown Brothers • 104 Culver Pictures

Chapter Four

111L Chicago Historical Society • 111R Loyola University Archives, Chicago, Ill. • 112, 113 Schenectady Museum Archives • 114, 126–127B Curt Teich Postcard Archives, Lake County (Ill.) Museum • 115 Library of Congress • 116 Chicago Historical Society (watercolor on cardboard by Charles Graham) • 119–125 Commonwealth Edison Company • 127T, 129, 134T, 137TL, 138, 139 Loyola University Archives, Chicago, Ill. • 128 Region History Archive, Lake County (Ill.) Museum • 130, 132 Chicago Historical Society • 133 courtesy of Forrest McDonald • 135, 137B *Chicago Tribune* • 136 Brown Brothers • 137TC, 137TR Chicago Historical Society • 140, 141, 146, 148 Tennessee Valley Authority • 142, 143, 145B Nebraska State Historical Society • 145T Franklin D. Roosevelt Library • 146 Antiochiana, Antioch College • 147 AP/Wide World Photos • 149 Daniel A. Miller • 153 *Time* magazine

Chapter Five

154–155 Center for the Analysis and Research of Spatial Information, Hunter College • 156–157 Port Authority of New York & New Jersey • 159R Cornelia Cotton Gallery • 159L, 160, 161, 162T, 165–170 Museum of the City of New York • 162B Library of Congress • 164 Museum of the City of New York/J. Clarence Davies Collection • 171 Chris Robinson • 172 Collection of the New-York Historical Society • 173–175, 177, 179, 186–187 Jervis Library, Rome, N.Y. • 176, 178, 180, 181, 183, 184B Cornelia Cotton Gallery • 184T Collection of the New-York Historical Society • 185 Museum of the City of New York

Chapter Six

188 courtesy of Margot Ammann Durrer • 189 Library of Congress • 190 courtesy of Allan Renz and Francesca Gebhardt • 191 The Carnegie Library of Pittsburgh • 192 Museum of the City of New York • 193, 194 *Scientific American* • 195 courtesy of Allan Renz and Francesca Gebhardt • 196 Collection of the New-York Historical Society • 197, 198, 200B Museum of the City of New York • 201 Library of Congress • 203, 207 *Scientific American* • 208 courtesy of Margot Ammann Durrer • 210 Tom Hilmer • 211 New Jersey State Archives • 213, 214, 215T, 216, 217, 219, 220

Port Authority of New York & New Jersey • 215B Collection of the New-York Historical Society • 221, 224 courtesy of Margot Ammann Durrer • 222, 223 MTA Bridges and Tunnels Special Archive • 225 Great Projects Film Company

Chapter Seven

226–227 Tom Way and Ginger Conly/IBM Corporation • 228–229, 231B Central Artery/Tunnel Project • 231T Central Artery/Tunnel Project (photo by Andy Ryan, 1990) • 232 Massachusetts State Transportation Library • 233T National Archives, courtesy of the Dwight D. Eisenhower Library • 233B Library of Congress (photo by Russell Lee) • 234 1978 GM Corp.; used with permission of GM Media Archive • 235 National Park Service, courtesy of the Dwight D. Eisenhower Library • 236 Master Highway Plan • 237T Massachusetts State Transportation Library • 237B, 238B, 241 Boston Public Library • 238T Michael Dukakis • 239, 250 Associated Press • 242 *Boston Phoenix* (photo by Peter Southwick) • 244 *Boston Phoenix* (photo by Jerry Berndt) • 246, 248 Central Artery/Tunnel Project • 251T *Boston Globe* • 251B *Boston Herald* • 253–256, 258, 259, 261–263, 267 Central Artery/Tunnel Project

Chapter Eight

269, 272B MIT Museum • 270 U.S. Army • 271T reprinted with permission of MIT Lincoln Laboratory, Lexington, Mass. • 271B, 272T The Computer Museum History Center • 273T photo courtesy of BBN Technologies • 273B Library of Congress • 276 RAND Corporations • 277TR The Computer Museum History Center/courtesy of IEEE • 277B courtesy of IEEE • 278 courtesy of Frank Heart • 279, 280, 281, 282, 289, 299 The Computer Museum History Center • 286–287 Brown Brothers • 290 Matisse Enzer • 292 Liaison by Getty Images • 294 MIT Museum • 300 The Computer History Museum/courtesy of the National Center for Supercomputing Applications (NCSA) and the Board of Trustees of the University of Illinois • 301 TIME Warner/AOL

INDEX

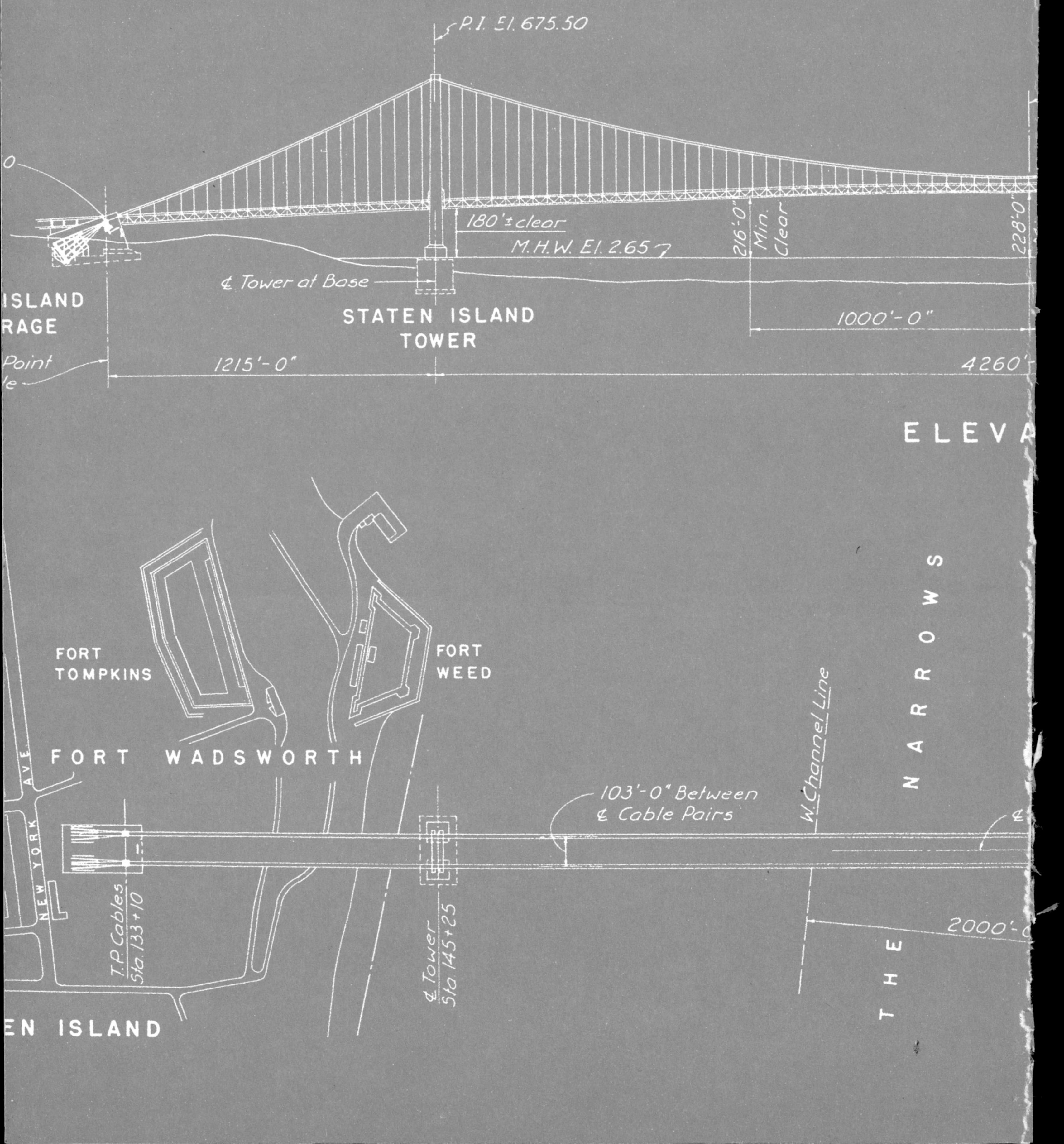

P.I. El. 675.50
180'± clear
M.H.W. El. 2.65
216'-0" Min. Clear
228'-0"
℄ Tower at Base
ISLAND
RAGE
STATEN ISLAND TOWER
1000'-0"
Point
1215'-0"
4260'
ELEVA
FORT TOMPKINS
FORT WEED
FORT WADSWORTH
NEW YORK AVE.
103'-0" Between ℄ Cable Pairs
W. Channel Line
THE NARROWS
T.P. Cables Sta. 133+10
℄ Tower Sta. 145+25
2000'-0
EN ISLAND